# TILL YOU FIND YOUR LOVE

Susan Deane travelled to Mozambique to try and find out more about her brother Peter's tragic death there. She just knew that it was out of character for him to have been involved in a common street brawl. On arrival, she made the long drive out to the *Quinta* Monteiro to see *Senhor* Raynor Monteiro, whom Peter had often talked about. Although Susan's meeting with this arrogant man turned out to be rather disappointing, it was to drastically alter the course of her life . . .

*Books by Barbara Best*
*Published by The House of Ulverscroft:*

THE HOUND OF TRURAN
WINGS OF DESTINY
ISLAND IN THE SUN
SECOND SPRING
A DREAM OF HER OWN
A MOST UNUSUAL MARRIAGE
TOMORROW IS OURS

BARBARA BEST

# TILL YOU FIND YOUR LOVE

## Complete and Unabridged

# ULVERSCROFT
Leicester

First Large Print Edition
published 1999

British Library CIP Data

Best, Barbara
   Till you find your love.—Large print ed.—
Ulverscroft large print series: romance
1. Love stories
2. Large type books
I. Title
823.9′14 [F]

ISBN 0–7089–4141–9

Published by
F. A. Thorpe (Publishing) Ltd.
Anstey, Leicestershire

Set by Words & Graphics Ltd.
Anstey, Leicestershire
Printed and bound in Great Britain by
T. J. International Ltd., Padstow, Cornwall

This book is printed on acid-free paper

# 1

'At last,' murmured Susan Deane with exasperation as she slid behind the wheel of her small car, throwing her wide-brimmed straw hat on to the seat next to her, adjusting her dark glasses. At last she could be off, driving carefully through the residential streets of Beira, away from this hot, bustling port on the Indian ocean, on to the road which would lead her into the interior, settling down for a long tiring trip.

She had hoped to make an earlier start, but when she arrived at the garage, a smiling Portuguese had informed her, with a shrug and an apologetic gesture of the hands, that the senhorita's car was not yet ready, his dark eyes saying, that which his lips dared not utter, how much he admired her fair, shining shoulder-length hair and dainty figure in its cotton frock and slip-on sandals.

Susan had abruptly turned away from that bold, embarrassing glance, pink banners of colour stealing into her cheeks.

Susan had docked the previous morning

and after having booked in a *pensao*, had visited the British Consular Mission. She had to have a good second-hand car and hoped they would be able to put her in touch with a reputable firm, as she had no wish to be stranded in the Mozambique bush.

Mr. Hardwick, a kindly man, with a mop of snow-white hair, had been a little dubious about her travelling alone, but had seemed satisfied when she told him that she could converse in Portuguese. She did not add, however, that her knowledge of the language was limited, for she had no intention of letting this fatherly man stop her plans.

Looking at him across his desk she added simply, 'I have looked after myself for years.'

He had smiled at that, with a keen look at her youthful figure and decided that she couldn't be more than twenty at the most.

'I'll arrange about the car myself, Miss Deane,' he had promised and gave her an address where she could call in the morning. I'll ring you this evening if I have been able to obtain a car for you, but I've had dealings with this firm before and have found them always reliable.'

'Please ask them to have it ready by eight,' she had stressed. 'It should take only three hours to reach Marula,' at which he had nodded. 'I'd like to get there before it gets too

2

hot and thank you for your help, Mr. Hardwick.'

Susan had poured over the route on the boat, tracing the outline of the main road as it wound up country, trying to memorize the names of the towns through which she would pass. In fact, the whole trip had been most carefully planned, right down to the last detail and now here she was on the last stage of her journey that, hopefully, would solve the riddle of her brother Peter's death.

The Suez Canal and down the east coast of Africa had been colourful and exciting, dancing practically every evening under a star-domed sky and sightseeing during the many stops at the various ports en route, but through all this gaiety, ran a thread of purpose. It had taken Susan a whole year of parred-to-the bone existence to save up for this trip, plus the little money she had had in the bank.

Her office job, which had been a good one, had not made her hoardings grow as fast as she would have liked, so had taken on an evening job as a cashier at a well-known restaurant in her area.

Susan eased her foot off the accelerator. The road had narrowed considerably and she had been warned by Mr. Hardwick, that when other traffic approached, or wanted to overtake, she would have to pull over to the side on to the gravel verge. Her thoughts went back to that job at the restaurant. Felix, her boyfriend hadn't liked the idea, at all and she grimaced ruefully. She had been hurt and upset when he had dropped out of her life. Why couldn't he have joined her? It could have been such fun; like Elma and Bill, two friends who also worked there. They had been quite happy washing dishes together, saving the money like misers for a longed-for holiday abroad.

The hot wind coming through the window of the small car ruffled her hair that was already growing damp at the temples, but Susan was unaware of the fact. There was so much to see that was strange and fascinating with each rise and bend of the road. For miles there would be nothing but bush and high grass, then it would open up to reveal a cluster of small houses with their attendant banana, mango and palm clumps; fields of pineapples, with their fruit atop reddish-yellow spikes, all signs of habitation, which was a welcome relief. Now she was passing through a dense forest area and met several huge trucks loaded with logs.

It was just over a year ago that the cable had arrived announcing that Peter Deane had been killed in a street fight in Marula. She had been aghast, but Peter's wife Nannette, had merely shrugged when the first shock had worn off.

'That's what comes of working in these God forsaken countries. Oh, the cities are lovely but the interiors!'

'Why did you ever marry him, Nannette?' Susan had asked curiously.

Her sister-in-law's big blue eyes had filmed with tears. 'I did love him,' she wailed, dabbing her cheeks with a wisp of lace. 'I did, but that trip with him to the Brazilian jungle was enough for me. You would have been able to cope, Susan, you're a survivor.' Her glance was resentful. 'The flies, the loneliness when the men were away from camp, the awful food and the heat!' She shuddered and wandered to the window to stare unseeingly at the rain curtaining the scene outside.

No, thought Susan, Nannette wasn't the type, a tight lump in her throat that seemed to be there permanently. 'You knew what his job was.'

There seemed to be no answer to that. Peter had come out here to Mozambique on another engineering project. Nannette had pleaded with him to stay in civilization and he had eventually promised, after many painful scenes, that when the current contract was finished, he would settle in England.

'Killed in a street fight,' the cable had said and Susan wondered unhappily if this would have occurred if Nannette had been willing to accompany him. Now it was too late. Ever since the cable had arrived, she had been determined to find out more about the incident. Her sister-in-law had written to the Portuguese authorities, but they had not been able to add to that terse cable.

'Well, I'm going to find out,' Susan had said grimly. 'I'm going out to Marula if it is the last thing I do.'

★ ★ ★

An hour and a half later she stopped again at a filling station, the junction where she would turn north and filled up with petrol, laboriously counting out the necessary escudos.

'Marula is only 64 kilometres,' replied the attendant in answer to her query, deciding that it was about forty miles further on.

'And the *Quinta* Monteiro?'

The man shrugged, a gesture she was beginning to know and distrust.

'Another eight or ten kilometres the other side,' he said vaguely.

That could be anything. Oh well and felt like shrugging herself as she got back into the car and turned on to the dusty, corrugated road. She glanced at her watch. Just after twelve, much later than she had bargained for. No wonder it was so hot and thought longingly of a cold shower and a change of clothing. It might be a good idea to book into a hotel in Marula before setting out for the *quinta*.

Peter had often spoken of the *Quinta Monteiro* and also about the man who owned it, one Raynor Monteiro and that was another reason why she had made this long trip, perhaps he could throw some light on the situation.

Had there been another reason? She could so easily have written, but this longing to visit this country, intrigued by the tales Peter had told her, had gripped her fancy. And Raynor Monteiro? Peter had painted him as a romantic figure, describing the safaris they had gone on together, roughing it in the bush, in such vivid detail, that it was no wonder that she had tried to picture this man. Peter had never spoken of a wife, just two sisters, both young who also lived at the *quinta*.

A disturbing thought crept in, making her tired eyes screw up even more. What if he knew nothing? But he must, thought Susan desperately. Her money wouldn't last long and her lack of knowledge of the Portuguese tongue would not allow her to apply for a job.

The miles slipped by slowly, the sun at its zenith, brilliant in a sapphire sky. Another twenty miles, she calculated, glancing at the mileage guage and blinked rapidly trying to clear her vision that had become blurred. Her back was stiff and a sharp, insistent pain had begun to stab at the base of her skull and across her eyes. Vast citrus orchards were on either side of the road and as far as the eye could see, but all her attention was focussed on the road in front as she tried to dodge the potholes that jarred her head every inch of the way, or so it seemed to the tired girl, splintering the pain into a thousand fragments and the dust billowed out behind her like a red, suffocating blanket.

At least she had the road to herself. It was the time of the *siesta* and that accounted for the lack of traffic. She had found out in Beira yesterday, was it only yesterday, that it was only tourists who roamed the streets, not willing to miss a single minute of their precious holiday. Most of the locals slept after lunch.

'*Mad dogs and Englishmen* — ' Well, it hadn't been her fault that the car had not been ready earlier, but admitted she hadn't made allowances for this awful heat, or this more awful road, either. She was crawling along, the car far too small to cope adequately with the corrugations and pot-holes.

Her head, by this time, was spinning alarmingly, making her feel sick and giddy. Oh, for a drink of cold water! What heaven that would be, licking her dry lips and wondered how far she would be able to continue under these conditions. A few miles further on made her decide to pull off to the side, opened her door, but found to her dismay, that there was no strength in her limbs and the pain in her temples jabbed like needles. Slowly she slipped back into her seat and thankfully laid her head on the steering-wheel.

If only the shrieking of the cicadas would stop.

Susan lost track of time, a strange lethargy creeping over her mind and body. She moved her head restlessly, barely able to open her eyes with the pain that engulfed her. Her visit to the *quinta* would have to be postponed. Suddenly she was still, her hands slipping from the wheel to lie dangling on either side of her limp body.

The sun had crept its long fingers well into the window by the time another car came down the road, stopping with a squeal of brakes as the occupant became aware of the slumped figure over the wheel.

Susan was vaguely aware of a figure at the window, trying to force her eyes open as a man opened the door.

'Are you all right?' asked a foreign voice sharply. 'You are in no fit state to drive.'

'I — I'm looking for the *Quinta* Monteiro,' her words only a whisper.

The man gave a start of surprise. 'Come then. It's only a few kilometres away.'

'No!' the word coming out with surprising strength, managing to raise her head.

There was amusement now in the voice. 'I'm Raynor Monteiro.'

Susan managed, somehow, to get out of the car and felt she had dwindled to pintsize. He was so tall, but when she tried to walk, her legs refused to carry her. With effortless ease the man picked her up and placed her carefully on the front seat of his car.

'Bonkers said he liked you,' she murmured, thankfully laying her head on the comfortable seat as the huge car sailed over the rutty road as if it had wings.

After that it was a paperchase of jumbled impressions, only vaguely aware of being lifted from the car and taken up a flight of stairs; knew the blessed relief of a tumbler of cold water held to her parched lips, which she drank with great, greedy gulps and then a cool pillow under her hot cheek.

It was hours later when she recovered consciousness again, to find a woman bending over her. She was middle-aged with thick black hair, liberally flecked with white, which was pulled severely back from a plump, wrinkled face. Suddenly recollection came to Susan, but when she tried to sit up, such a fierce pain shot through her temples that she was forced to lie back again.

This must be the *Quinta* Monteiro.

'Please to keep still,' admonished the woman firmly, then addded, 'You like a drink, no?' in broken English.

'But — but what happened,' asked Susan. Her Portuguese as bad as the woman's English and was surprised how hollow and weak her voice sounded. It was more like a croak and her lips felt cracked.

'You speak our language, *senhorita*,' and the woman broke into a flood of words.

'*Senhora*, I only know a little and can only understand if you speak slowly.'

'You seeck, too mooch sun and fever, yes,' and held a glass of cool orange juice to Susan's lips. 'And now you sleep.'

Susan obediently did as she was told, but only until she heard the door close softly, then looked around her. Pretty cotton

11

curtains fluttered in the slight breeze coming through the big window and the snowy-white net curtains filtered out the glare from outside.

There was a heavy, carved oak dressing table and wardrobe, which she recognised as being good and were probably very old. One only occasionally saw workmanship like this now-a-days. She closed her eyes now that her curiosity was satisfied as to her immediate surroundings, then tried to picture the man who had rescued her, but tantalizingly, the memory eluded her. The only remaining impression was that of strong arms that had picked her up as if she had been a child, but what did he look like? What she did remember was his very good English, with only a trace of an accent. On that memory sleep took her again, as easily as a drifting leaf in an autumn breeze.

When Susan woke again, she felt much better, although her body still ached all over, her head was clear, but there remained a lethargy that made her quite happy to stay where she was for a little while longer.

A small person tiptoed into the room, a hot little hand thrust into hers.

'Does the head still ache badly?' came the solicitious enquiry.

Susan gazed drowsily at the little girl. '*Bom*

*dia*,' she murmured. 'Are you my guardian angel?'

'Oh no!' came the giggled reply. 'I'm not an angel. *Tia* Maria says sometimes I'm an insect.'

'Oh,' breathed Susan delightedly, a chuckle escaping her.

'Have I said the wrong word?' asked the child anxiously.

'I think you must mean a pest and I expect she called you that when you were bothering her.'

'Yes, I expect that was it. You understand so nicely. I do hope you will be able to stay here for a very long time.'

'Your English is excellent. Do you learn at school?'

'Raynor say English is important, so we must learn.' This was said with an old-fashioned air of wisdom. 'We do have Miss Brack, but she has gone to visit her parents, so we're having a holday, but I do go to school to learn my own language.'

'And who is *Tia* Maria? Is she your sister?'

Another giggle broke from the little girl accompanied by a violent shake of the head. 'Oh no, she's a way-back relation, you understand, and looks after us and the house. *Tia* means aunt and my name is Falina and I'm seven years old and now please, how are

13

you feeling?' All this rushed out without a break, her dark brown eyes nearly level with Susan's, round with interest. Dark hair, cut in a straight, uncompromising way below the ears and across the brows gave the illusion of chubbiness to the small face.

'I'm feeling much better, thank you. It must have been the heat that made me faint. It was dreadfully hot yesterday in the car. I felt just like the Sunday joint sizzling in an oven.'

Falina chuckled then clicked her tongue sympathetically. 'It wasn't yesterday you came to us,' with a violent shake of her head, that made the heavy hair swing across her cheeks. 'It was three days ago and I've been hoping and hoping for you to wake up and talk to me. You looked so pretty lying there with your golden hair so soft and curly and oh, you have blue eyes, just like the story — '

'Three days!' Susan gasped. 'Have I been here that long?'

Another nod. '*Tia Maria* put you to bed after Raynor had carried you up here, just like in the story. Oh!' Falina's hands rushed to her mouth as if she had just remembered something. Slipping off the stool, she rushed out of the room, but was soon back accompanied by the woman she had first seen. This then must be *Tia Maria*, the way back relation.

'So you are awake,' she said, laying a hand on Susan's forehead. 'That's good.'

'*Senhora!*' There was real perturbance in Susan's voice. 'I'm so sorry!' but her apologies were waved aside.

'The *Marquês* asked to be excused for not seeing you, but he was called away soon after you were taken ill. Are you a little recovered?' she added politely, then turned to Falina to ask her to refill the water jug.

So he is a *Marquês*, mused Susan. What a lot she would be able to tell Nannette when she wrote. 'I'd like to get up, I can't trouble you any further. I'll go to the hotel.'

'No, no, the doctor he is coming this morning. Better you wait until he comes. Besides the *fidalgo* said you stay, to be looked after until he come back,' and started to tidy the pink coverlet. 'Do you have business with the *Marquês*?' she asked curiously at last.

Susan looked a little uncertain. 'It's really not business — ' she began.

*Tia Maria* nodded, as if satisfied about something, suspicion now in the small, black beady eyes. 'It was fortunate that the *senhorita* was taken ill so near the *Quinta*,' she said with marked coolness.

Susan protested hotly. 'It was not my fault

15

that I was taken ill.'

The fat black clothed shoulders shrugged, the short fingers splayed in a disbelieving gesture, but a knock at the door prevented her from replying. She hurriedly went to the door and opened it.

'*Bom dia*, Dr. Martenis,' she beamed and introduced him to his patient.

He was a round, squat little man, with a balding head ringed by a fringe of greying hair and a harassed expression on his red face. He looked as if he had far too much work to do. This proved to be true for Susan found out later that he attended a radius of some fifty miles, but his smile was warm.

'Much better this morning, I see,' he said with satisfaction, lowering himself gingerly on to the stool that Falina had used, his paunch resting comfortably on his knees.

'May she get up now, Dr. Martenis?' asked Falina eagerly. She had come in with the doctor and was now standing beside him. 'I want to show her the garden and the house and everything.'

The doctor had been observing Susan keenly under cliff-like brows. 'Does the head still trouble you?'

'No, and that awful dizziness has gone too, thank goodness,' shaking her head to prove it.

'*Muito bem*, she can get up, *Senhora*

Ferreira,' his fingers still on Susan's pulse, 'Just for a while today, though. We'll take things quietly.'

'*Muito obrigado*,' said Susan shyly. 'Everyone has been most kind.' How very catching was this pedantic way of speaking, she thought wryly and what was more, her accent was all wrong.

As soon as the doctor had left, *Tia Maria* brought Susan her suitcase. 'I have only hung up your dresses. The bathroom is through that door,' and with that she shooed Falina out of the room, but not before the little girl had called out that she would be back.

Susan showered but found to her dismay that her legs were strangely wobbly and was thankful to be able to flop down on to the bedroom chair. Even brushing her hair had been an effort.

The next few days went by slowly as she was only allowed up for a few hours each morning and afternoon and although she did regain some strength, she began to worry. Her precious time here in Mozambique was fast running out and she was eager to talk to the *Marquês*, but he was still away. The following morning she was sitting on the balcony, enjoying the lovely view of the garden below, when Falina came to fetch her.

'Leonora — that's my sister,' she explained

hastily noticing Susan's look of enquiry, 'would like to meet you and *Tia Maria* says a little walk will do you good. Leonora isn't very well, you understand. Just lies on her bed all day.' The little girl shook her head mournfully.

She took Susan along a wide, carpeted passage and knocked on the door.

It was the most feminine room Susan had ever seen. A quick glance showed her pale-pink satin curtains over white nylon drapes gently blowing at the wide french doors that led on to a balcony and a white wall-to-wall carpet. What luxury, and then her gaze was taken up by the girl who lounged on a day couch. Her compassion was instantly aroused.

The oval face, with its sallow complexion, held a sullen expression and the long black hair hung down, dull and lifeless.

'This is Leonora, Susan,' said Falina.

'*Bom dia.*'

'I'm sorry to hear you have been ill,' said the girl waving her hand to a chair.

'I'm much better, thanks. And you, *senhorita?*'

'Oh, I'm — I'm not ever really well,' admitted Leonora in a voice that would have been bored but for the painful stammer. A flicker of interest, however, filled her dark eyes, for a moment, staring at Susan's fair, curly hair. 'Is — is that real?'

'Yes, but it is the disappointment of my life. I want straight hair, like yours.' She chuckled suddenly. 'I even got my mother to iron it, would you believe that?'

'Ironed it!' both girls chorused.

Susan nodded. 'I just laid my head on the board for my mother to iron. My hair was straight for a few hours,' and was pleased to see Leonora's grin.

'You must have felt like all those queens who laid their heads upon a block.'

'So you know a little English history?'

'It's Miss Brack's favourite subject. Tell me about things you are alllowed to do,' only polite interest in the slightly accenuated words.

'There isn't very much to tell. I lived in a flat in London and worked as a shorthand typist.'

'You lived with your parents, or perhaps a companion?'

Susan shook her head.

'I lived by myself. My parents were killed in a plane crash some years ago.'

A small, warm hand was slipped into hers. 'We also lost our parents, but not together, but Raynor looks after us, only he is away so often.' Falina gave a tiny sigh of regret, as she lent against Susan's shoulder.

'Are you allowed boyfriends?' queried Leonora sitting up suddenly, swinging her legs off the couch.

Before Susan could reply she went on. 'Tell me about them,' an animated gleam to her eyes. 'I have heard that English girls can do what they like. Is it true?' Susan blinked at the change of mood.

'Well, we don't have chaperones, if that's what you mean and I often went out with boys; mostly with a crowd though.'

'Oh!' she sighed enviously, sinking back once more on the couch. 'How wonderful to be you! To do exactly as you like. If I did that, Raynor would drop dead immediately. You say there is nothing to tell in your life, but — but what about — mine?' a mutinous curve to her soft full lips. 'I can't — do this, I — I can't do that and — and what is more I'm not yet affianced. I don't think — that my brother has even made an-y arrangements.'

This was said with such an air of grievance that Susan gave her a keener glance, wondering how old the girl was.

'But you have been so ill,' cut in Falina soothingly. 'When you are better, perhaps — '

Leonora shrugged pettishly. 'But what is there to do here? If only I was allowed a boyfriend. You don't know what it is like to be born Portuguese.'

A bell rang through the house and Susan glanced enquiringly at the girls.

'Lunch time,' said Falina. 'Come, Susan, I'll show you the diningroom.'

'Tell *Tia* I'll have a tray up here,' said Leonora, lapsing into lethargy again.

Susan stood at the top of the wide stone staircase with its black balustrade of wrought-iron and gazed down with awe at the huge hall below. The wood floor was highly polished and the heavy oak door looked as if it could repel an invasion and had probably done so. If this was the hall, what was the rest of this mansion like? Falina took her through the hall out on to a verandah that encircled wholly an inner courtyard.

Here was a garden, but what a garden! Susan's eyes widened, but she was only allowed a glimpse of exotic flowers and a cool green lawn before being led into the diningroom which was furnished with heavily carved oak that had a patina of loving care.

After lunch Susan was quite happy to get back on to her bed with a great deal to think over. Would she soon meet the *Marquês* and then there was Leonora. She grinned suddenly and was quite sure Leonora would not have envied her quite so much if she had been obliged to live in that bed-sitter she had had to put up with, not that she had ever minded, appreciative of having a place of her own, but what was wrong with the Portuguese girl? Was it some rare tropical disease?

Susan snuggled drowsily deeper into the cool pillow and wondered how she was going to be able to drive back to Beira feeling like this, but fell into a deep sleep before she could worry any more about either question. She was awake when Falina popped her head around the door, feeling much refreshed.

The little girl led her to a small sitting-room at the end of the passage.

'Please sit down and I'll tell Timbo to bring up the tea. *Tia* won't be long,' Susan very aware again of the child's very good manners, but it was not Falina or the *senhora* who accompanied the servant with the tray, but a tall, lithe man immaculately dressed in a light-weight grey suit, his dark hair brushed back from a thin sardonic face.

This must be the *Marquês*. He couldn't possibly be anyone else, for he had that assured manner of a man in his own domain. His light brown eyes were unexpectedly penetrating. There was also about his mouth a mocking smile that Susan viewed with misgiving. Did he also think, as the old

22

*senhora* had done, that she had purposely been taken ill near the *quinta*? Her cheeks paled as she tried, unsuccessfully, to stop the flurry of apprehension that was stifling her throat.

# 2

Susan squared her shoulders as the *Marquês*
sat down opposite and came to the
conclusion that she was not going to like her
host and his mocking appraisal, that sent the
colour wildly, protestingly to her cheeks, was
beastly.

Timbo handed the tea around accompanied by a plate of homemade sweet biscuits
and then disappeared.

'I'm pleased to see you are so much better,'
he said politely. 'I must apologise for my
absence. I hope you were well looked after?'

'Yes — yes thank you, s -senhor,' and
found herself, much to her dismay, stammering nearly as badly as did Leonora.

'And now, please, I would like you to
explain who this *Senhor* Bonkers is you spoke
of. I've searched my memory all these days,
but nothing comes. I'm sure I know no one
by that name. Enlighten me.'

'I'm sorry, but I don't understand.'

'You probably don't remember, but the last
thing you said before losing consciousness
was this — this person liked me.'

A nervous urge to giggle nearly choked her.

What had possessed her to use Peter's old nickname? She hadn't thought of it for years. '*Senhor*, I have come to Mozambique to find out more about Peter Deane's death. Bonkers was his nickname,' she explained hastily. 'I'm sorry to have caused all this trouble, but I must find out what happened that night.'

The sudden frozen look on the *Marquês* face made her falter and once again gazed at this man in utter bewilderment. He had turned away, his manner haughty and withdrawn, realising how cold brown eyes could really be.

'Peter Deane was killed over a year ago. Isn't it rather a long time to have waited before making enquiries?' His tone dripped icicles. 'And why come now, *senhora*?' Well, his lifted eyebrows demanded. 'Why could you not have come out here with him and made a home for him? Peter was a lonely man and what is a wife for, but to comfort her husband, where ever he goes?'

Susan had shrunk back into her chair, but there was no escaping those dark, intense eyes so filled with contempt. His words had cracked out like lashes from a whip.

'But, *senhor* — ' she protested, with all the dignity her twenty years could muster, through lips that had gone dry, 'I was not his wife, I'm his sister.'

Anger came to her aid, driving out fear. This was not the man who had run so persistently through her thoughts, the romantic figure woven from Peter's letters. She should have known, Susan reflected bitterly, that dreams never came true. This man was nothing but an arrogant despot, trading on his title and vast estates.

Her wide deep blue eyes glared back at him as she straightened again in her chair, but just as she was about to pour out hot words, his anger broke.

He leaned forward apologetically, that dark glance flashing over her face. 'Senhorita, a thousand pardons! What I presumed was unforgivable. The sister of Peter Deane is most welcome here at the Quinta Monteiro.' A gleam of humour softened the lean lines of his face as he went on to gently chide her. 'I do think, perhaps, you could have introduced yourself sooner and this stupid mistake of mine could have been avoided. It was a natural mistake on my part to think you were Peter's wife when you said you were making enquiries about him.'

His voice was foreign and dangerously pleasant and she shied away from it with suspicion.

'But didn't you know who I was?'

'No, I came straight up here only meeting

Timbo with the teatray.'

So he hadn't yet seen the old *senhora*. 'Didn't you open my bag to look for my passport, or any letters?'

The *Marquês* shrugged with distaste. 'I thought it time enough for you to tell us who you were when you recovered.'

*Tia* Maria had had no such scruples, for she remembered the *senhora* mentioning her name to the doctor at his first visit.

'Now tell me what I may do for you,' he said, getting up to pour their second cups of tea, that had lain forgotten between them. Again the slight inclination of the head, but Susan refused to be mollified.

'The cable only said that Peter had been killed in a street fight.' Her head went up proudly, colour tinging her pale skin. 'I do not believe that, *senhor*. My brother wasn't the type to go fighting in the streets.' She glanced at him quickly, then away again, the long lashes coming down over the troubled eyes. 'Was he so lonely that he took to drink out here?' hesitantly voicing a fear that had gripped her ever since that cable had arrived. 'That is what has bothered me and yet, another part of me knows he would not do a thing like that, either.'

'Peter was lonely, yes *senhorita*, but he never drank to excess. I can at least put your mind to rest there.'

Susan's face glowed, her eyes deep wells of relief, as she took the cup handed to her.

So, this English flower could come alive quite delightfully, thought the *Marquês* and gave her a keener glance. 'I'm curious why his wife didn't come out with him in the first place. It is also quite incomprehensible that a wife should not want to know what had happened to her husband. Why did she not accompany you?'

'Nannette is not cut out to rough it, *senhor*. I doubt very much if she can even fry an egg.' As she saw a frown pucker his brow, added hastily, 'She did go out to Brazil with him and stayed there quite a while.'

'With an abundance of labour in this land, *Senhora* Deane would not even have had to fry an egg,' he retorted evenly. He leant over and placed his cup on the beautifully carved table. 'Couldn't the authorities help you?'

Susan shook her head. 'I was hoping you would know something more. Peter spoke so much about you and I couldn't come sooner,' indignation welling up again as she remembered the other accusation that had been levelled at her, even though he had thought she was Nannette. 'I had to save for the fare.'

'You want to clear your brother's name, is

that it?' his glance warmly sympathic, 'but it is most unfortunate that I too know no more than you do, *senhorita*.' The gesture of his long fingered hands was one of regret.

Dismay was written deep on Susan's face. She had counted on this man to give her other information. 'I'm due to sail back home early next week. I'd better try elsewhere. The construction camp, perhaps?' wearily passing a hand over her eyes, the depressing tiredness once more descending upon her. It would be too dreadful if her journey was to be all for nothing. 'I'll book in at the *pousada* in Marula. Thank you, *senhor*, for all you have done for me and now, please excuse me, I must go and pack.'

The mocking smile was back. 'Tch! You independent English and why the hurry? Can you not accept my hospitality for a few more days?' he asked smoothly, 'or is it you do not wish to accept it from a foreigner?' His eyes had begun to smoulder again, his lips a taut line.

'But, *senhor*, I can't — '

'Can't Miss Deane? That is a word not in my vocabulary.'

She stiffened, she could well believe that.

'I insist you stay here,' went on the *Marquês*. 'As a friend of your brother, I think I have the right and I'm sure he would not allow you to travel around this country by yourself.'

How insufferable he was, she thought as he waited for her reply. Indignantly she met his glance that, for once, wasn't at all cynical. He was only extending this invitation from a sense of loyalty to Peter, she decided and rose purposefully from her chair. 'Thank you, *senhor*, but I feel I must go. I have already presumed too long on your hospitality.'

The *Marquês'* lean figure tautened and then, with an angry movement of his hand, strode to the window, signalling her to be seated.

'Would the offer of a position here at the *quinta* make you stay?' he asked abruptly.

'But is there a position?' She hedged a little at this unexpected suggestion.

'You could take Miss Brack's place.'

'But isn't she coming back soon?'

'I've had a letter to say her parents are ailing and she would like to stay a few more months with them. Well?' He turned from the window. 'Would that appease the obstinate streak of pride that does not allow you to accept my hospitality?' He was openly mocking again.

'But what about my job in London and my booking back and — and the *Senhora*

Ferreira?' This was tacked on as an after thought for she was very sure *Tia* Maria would not welcome her here as a companion to the two girls. She had already shown her dislike of Susan.

'*Tia* Maria, what of her?' He shrugged his shoulders. 'As for your other excuses — ' snapping his fingers, 'they are as nothing. I'll see to them.'

Susan could well imagine the letter the *Marquês* would write to Mr. Manton, her boss. A letter that would leave him in no doubt that he was of no consequence whatsoever. The thought made her smile, but there was no getting away from the fact that the arrangement would suit her very well and would give her extra time in which to meet other people who had known Peter.

Susan knew that she would be more of a companion to the two girls, as it would not be necessary to teach Falina, as the little girl went to school in Marula and yet she still hesitated to accept the *Marquês*' offer, an uncanny feeling taking possession of her, a warning that if she stayed, this man's impact on her life would affect her deeply. Whether for good or ill, she had no means of telling.

'I'm waiting for your answer, Miss Deane,' and then in a gentler voice and a warm smile added. 'I think you'll be good for Leonora.

*Tia* Maria and Miss Brack are getting on in years. As for Falina, she already loves you deeply. I would not like to hurt or disappoint that little one. I met her on the stairs as I came up and she was worrying about you having to leave.'

It wasn't fair bringing in Falina, thought Susan, for she had also grown very fond of the little girl and found herself agreeing to the *Marquês'* suggestion.

'*Bem*!' he said, but with a satisfaction that somehow gave the impression that the outcome was a foregone conclusion anyway. 'If you'll let me have the name of the firm you worked for and your return ticket, I'll see to everything.' The interview was over.

Falina was hysterical with joy when she learned the news that Susan would be staying and even Leonora showed unexpected interest. Here was a matter that needed her immediate attention, thought Susan, deciding to have a chat with Dr. Martenis at the earliest opportunity. If she knew something about her medical history, perhaps she would find a way to help this Portuguese girl. The doctor had seemed an understanding sort of person.

★  ★  ★

Susan found her duties light. She saw Falina off to school at seven thirty each morning, after having supervised her dressing and breakfast, then arranged the flowers in the main rooms downstairs, delighting in picking huge armfuls of pointsettias, their big, velvety heads blazing colour in the darkest corners. The beautiful strelitzias with sword ferns filled the open fireplace in the *sala*, while the perfume of pale yellow frangipani followed you up the wide staircase.

She had become quickly accustomed to this unusual mansion, Falina having taken her around one afternoon and had proudly shown her everything. They had started from the inner courtyard where they had been sitting. It was deliciously cool here, surrounded by a wide variety of flowers and shrubs that grew in wild profusion, with ferns cascading out of tubs and hanging baskets, in curling waves of green. Camelias, magnolias and begonias seemed commonplace, except that they were the biggest Susan had ever seen. This mansion was more like a castle, thought Susan, as she had gazed about her, but there was no feeling of being hemmed in. Falina had told her that the house had once been a fort, a very long time ago.

From there they had done a tour of the house. There were only two outside doors and open verandahs on each floor ran right around the courtyard, the rooms opening out on to it.

'Now come and see the ballroom,' Falina had said and took her into an enormous room. 'There hasn't been a dance here for years,' she had added regretfully, 'but perhaps for Leonora's eighteen birthday?'

Susan had looked around her with delight. The mosaic floor, the chandeliers, the magnificent murals, the heavy blue and gold velvet drapes at the high windows, all denoted an air that surely did not belong to this modern age. Here was evidence that all this had been lovingly preserved over the years. She had never dreamed of ever being tippled into such luxury.

Next, on the south side, was the *sala* where the family entertained their friends and also the diningroom. Facing north was the suite belonging to the *Marquês*.

Susan only peeped into the room, quick to notice its monklike bareness. None of the sumptuousness for him, evidently. What kind of a man was he really, this employer of hers, she wondered curiously. Upstairs were all the bedrooms.

Falina was brought home at one o'clock in the chauffeur driven car, looking small, but very important sitting on the wide back seat by herself. The midday meal was a family one, although they seldom saw the *Marquês*. After lunch everyone in the *quinta* had a *siesta* until four o'clock when tea was served. It was like the enchanted castle of the fairy-table, where everyone slept, from the king to the humblest scullery maid.

After tea, Susan gave Falina an English lesson and then when that was over, they played games, on the strict understanding that only English was spoken. Susan and Falina had their supper upstairs, Leonora sometimes joining them, other times she would make an effort to dine downstairs with her brother and the old *senhora*. More often than not, she had a tray brought up to her room.

It was all so unnatural, thought Susan and the system did not help either, for all Leonora had to do was press a button and a servant came running.

Susan was shocked that a maid still bathed Falina and put out her clothes for the following morning. A child of this age should do this for herself. Falina had been delighted with the suggestion that she do things for herself and the maid had gone off quite happily to have a gossip with her cronies in the kitchen, a wide smile splitting her dark face, showing the gleaming white teeth to good effect.

Susan had sat on the edge of the bath, supervising and watched the child's utter absorption. Then there was a time for a story before the supper trays were brought up and by eight o'clock Falina was in bed.

And that is my day, thought Susan as she sat in an easy chair reading in the upstairs lounge. She had peeped in at Leonora to find her curled up on her bed, a book in her hand and had asked her if she would care to join her, but Leonora had declined. Susan had not felt she could sit down on the couch without an invitation and sensed she would have to feel her way around a lot more before she finally won her confidence.

★　★　★

The sun was pleasantly hot when Susan and Falina set out the next morning, in Susan's small car, for Marula, to do some shopping for the *senhora*. Now that Miss Deane was a paid employee at the *quinta*, she must be made to be of some use. Wool to match and groceries to order, a message to take to *Senhora* Tomaas and would the *senhorita* please not forget to take this little jar of honey to Dr. Martenis. This suited Susan for she wanted to see the doctor and parked under a tree in the now familiar *Avenida* Monteiro.

The small town buzzed with activity. White clad cookboys, with enormous sisal baskets, chattered noisily at the open air vegetable and fruit stalls, while the men lazed at a café that spilled over the pavement, sipping coffee under gay striped umbrellas.

Susan tried to avoid the interested glances thrown at her and kept her eyes turned towards the shop windows which, though few, had a surprisingly large variety of wares that had been imported from the mother country, Portugal and also from the provinces.

The houses were old and solid, built with the local grey stone, square and almost forbidding in design, except that they all had verandah boxes bubbling with colour.

'Everyone forgets to say '*Bom dia*' to me,' chuckled Falina, holding Susan's hand. 'All they can do is look at you. Nobody in this town has fair hair and blue eyes,' peeping up roguishly at Susan's flushed face. 'And nobody here can blush either.'

The shopping finished at last, they drove to the hospital. The building, a low sprawling whitewashed structure, looked as if it had been added on to many times over the years. Susan found Dr. Martenis in the Outpatients Department.

The little doctor bowed gallantly as he

ushered his visitors into his office and then, with a side glance at the little girl said, 'Come with me, *pequena*. We'll see what Sister has for you.'

He was soon back and pulled out a chair for Susan, thanking her for the honey. 'Now, Miss Deane, how may I be of service to you? Is the head still troubling?' sliding adroitly into the creaking swivel chair behind his desk. It sagged alarmingly under his weight.

'I feel fine, thank you. It's Leonora.' She hesitated. It had seemed so right to come to this old man, but as she faced him across his littered desk, wondered if she had been too hasty. 'I don't know if I should first discuss this matter with the *senhor*. I have taken Miss Brack's place until she comes back.'

'What is your problem?'

'I would like to help Leonora, if I possibly can. Doctor, why does she just lie there? It's so unnatural. Is she ill?'

The doctor shook his balding head, fiddling a pencil between his short, stubby fingers, his gaze on the jar of honey as if it had been a crystal ball.

'She broke her leg very badly in a riding accident some six — ' he shrugged, 'seven months ago. That has healed completely, but her leg is, naturally, still weak because she does not do the exercises I have given her.

Other than that, there is nothing wrong with her. *Com a breca!* it is terrible for me too, to see her just lying there,' pulling his mouth down and waving his hands expressively. 'There is nothing more I can do. If she had been born into a less privileged family, she would have had to exert herself. As it is — ' He shrugged resignedly again.

It must be something else then. Was it her stammer, but Susan had noticed that when the girl was angry the stammer disappeared. Was this the way Leonora faced reality, she wondered, shutting herself in her room, afraid to meet people because of this impediment, plus the fact that her brother had done nothing about arranging a marriage for her? She had spoken so bitterly about that. Did she now consider herself a social outcast and found the easiest way out was to pretend to be a sickly recluse? Susan's heart ached for her.

'*Muito abrigado, senhor,*' Susan said getting up from her chair. 'I won't keep you any longer.'

'You have an idea, no?' his glance keen under the bushy eyebrows.

'I'll have to think about it a little more,' smiling apologetically at the doctor as he opened the door for her.

He beamed and shrugged his massive

shoulders. 'I understand. Someone new — who knows, *senhorita?*' and went off down the passage to fetch Falina.

'*Adeus, senhorita,* Falina. I will be most interested in the outcome of your deliberations, Miss Deane,' and smiled, shielding his bald head from the sun with one hand while opening the car door with the other.

Susan mused on the problem all the way to the *quinta.* It was very true what the doctor had said. If Leonora had been born into a peasant family there would have been none of this isolating nonsense. She would have to talk to the *Marquês* after all.

As she drove along the avenue of flame trees to the house, she found the prospect chilling, but if Leonora was to be helped, there was no other alternative.

After lunch Susan settled Falina for her rest and as she did not feel the least bit tired, decided to write to Nannette, a task she had been putting off for days. It was very pleasant on the balcony, but the garden called and gathering up her writing material, she slipped through her room, down the stairs, settling herself on a wide seat in the shade of a wild fig tree.

All the colours here hurt, she thought gazing at the brilliant flowers, none in pastel shades, colours so intense that they appeared unnatural. Hibiscus, bouganvillea and ginger-bush flaunted their finery shamelessly in the hot sun. The hoses were running and at the far side of the large lawn, a sprinkler sparkled a rainbow in a wide arc.

Susan opened her writing case reluctantly and wondered if Nannette would really be interested in her news. A month before she had left for Beira, her sister-in-law had just arrived back from the Continent with a trunkful of lovely clothes and the evening Susan had visited her, was as fragile as a flower in a filmy black lace frock that had enhanced her fair skin and the blue of her eyes. Rumour already had it that Nannette was going steady with a man who was something big in the Stock Exchange world.

It shouldn't have been such a shock, for Susan had never thought or expected Nannette not to remarry, but nevertheless, it had been a jolt.

Susan unscrewed the top of her pen. She had to let Nannette know that she was staying on in Mozambique and about the offer she had received and about the *quinta* with its beautiful antique furniture; about the small, brilliant, quite fascinating honeybird that had just flashed down to perch on a redhot poker, sipping nectar from its many chalices, but she said nothing about the

41

*Marquês*, except that Raynor Monteiro had been very kind.

Susan's gaze shifted to the north wall of this huge house, to its battlements, its wrought-iron grilles and balconies. It appeared to have been bedded down here for hundreds of years, glistening white in the afternoon sunshine and she knew she hadn't really done it justice in her letter.

Why had she omitted the important fact that its owner was a *marquês*? Susan turned quickly as the huge black car belonging to her employer and which he alone drove, swept up the avenue and purred to a stop in front of the house. As he came in to the hall he saw her and crossed the lawn.

'*Boa tarde, senhorita*,' he said with a bow. 'I thought it might be you and why are you not resting like the other members of my family and household?'

'You don't take a *siesta, senhor*,' she reminded him with a smile.

'True,' he conceded, slipping into the chair next to Susan, 'but I would remind you, Miss Deane, that I've lived in this country all my life, except for the month each year when I visit Portugal. Besides, I have much to do. Well, Miss Deane?'

'And I'd much rather sit in this beautiful garden, instead of wasting time asleep upstairs,' she retorted. 'It's lovely and cool here.'

'You do not want a reoccurence of heat exhaustion, *senhorita*. The good doctor has warned that you will have to exercise much care in the future.'

'But I'm better now — ' she protested.

'I'm very glad to hear that, but I think you'll not miss the *siesta*.' The words so smoothly spoken had an underlying firmness that could not be ignored. He must have seen her rebellion, for he went on gently. 'This is a wonderful country, but for some the heat is like a poison and so they have to leave. I should not like that to happen to you.'

'But, *senhor*, I love the heat,' Susan assured him quickly.

'That is good, but you'll continue to be careful.' He turned towards the house. 'You are interested in my home?'

The subject had been changed and she did not dare go back to it, but she fumed inwardly. Her freedom was being curtailed, even to the extent of laying down the law about the *siesta*. It was too much! Her eyes were smouldering as she replied. 'It is a most unusual design, *senhor*. Was it really a fort at one time?'

'Yes, a hundred and fifty years ago. It was

built by the first Monteiro to settle in this country. He had much land and brought many workers out from Portugal. The country was unsettled then and so he had this *forte* built where everyone could come in times of trouble, even the local inhabitants. It was my grandfather who had it converted to its present design and who laid out the gardens.'

The *Marquês* drew out an expensive cigarette case from his pocket and offered one to Susan who refused. 'You are happy here with us, Miss Deane?' he asked, bending his head to the flame as he flicked open his lighter and when she nodded added, 'I like to think all my people are contented.'

Another gust of annoyance shook Susan. So that was how he thought of her, just another of his vassals. How feudal!

'I do not belong to you, *senhor*,' she replied with a cool little smile, refusing to bow to his arrogance.

Up went a dark eyebrow. 'I pay you and your welfare is very much my concern, *senhorita*.'

She bit back a sharp retort. Why did this man always annoy her so? Was it the foreign turn of phrase, she wondered crossly and then remembered that there were certain things to discuss with him. 'I would like to talk to you about Leonora, if you can spare the time?'

'Yes, Miss Deane?'

He had half turned to her, his arm along the back of his chair, his lean face quizzical. It wasn't easy, with him in this mocking mood but at least he was affable this afternoon and so she plunged on. 'I hope you don't mind, but I spoke to Doctor Martenis about your sister. He cannot understand either, why a healthy girl should lie about all day, as Leonora does.' Her glance was steady.

He nodded. 'Please continue. I'm interested to hear your opinion.'

'*Senhor*, I've been told that this stammer of hers has only appeared since her accident. I think Leonora is hiding because of it.'

'A psychological link up?' She nodded and he went on, 'Have you a suggestion that might help?' His eyes were keen as he looked at the earnest young face across the garden table from him, with the bloom of an English peach.

'She could take up singing,' she suggested. 'It would correct her breathing and show her how to control her words.'

'That might be a problem — '

'No, I could teach her.'

'So, you sing?' he asked politely.

She shook her head, unaware that a small

glint of mischief had shown in her eyes. 'I was a great trial to my parents who were both music mad. My father was the conductor of the municipal orchestra of our small town and Mum played the piano. I was taught music but I learnt the technique, nothing more.'

'But, Miss Deane, I think there is something else on you mind, not so?' he asked astutely. 'You could have done this without my permission.'

She looked at the *Marquês* steadily. 'Is it possible for you to spend more time with your sisters? Take them about with you. Leonora is nearly eighteen.' Her tone was accusing.

'Ah, ha! Now we come to the crux of the matter. I'm very attached to my sisters, but I'm also a busy man.'

'I realize that, *senhor*,' the reply came quickly, 'but Leonora badly needs to be taken out of herself, made to feel part of your family, an important member.'

'You have not been here long enough to know that we visit many families. Leonora will not be persuaded to join us, however. What more can I do?' he demanded impatiently.

'But these people she has, probably, known all her life. What is needed are new interests,

46

something to jolt her out of her present rut.'

The *Marquês* shrugged and put out his cigarette. 'So now you blame me for Leonora's defects?' Anger bubbled up again between them.

'No, *senhor*,' her breath coming a little faster, 'but surely you could take Falina and Leonora with you when you visit the allotments and fruit groves, or perhaps take Leonora to Beira when you go there? Would you do the same to your wife, leaving her by herself day in and day out, as you leave your sisters? You have a duty to them.'

Her employer stiffened and Susan knew she had gone too far. There was a carved look about him as he said curtly, 'You forget yourself, Miss Deane.'

'Yes, I know I should mind my own business,' she flared back at him, throwing caution to the winds, 'but Leonora and Falina are also my concern. When I take on a job I do it to the best of my ability. If I didn't it would be cheating.' Her eyes, the colour of the heavens, sparkled angrily across at him. 'And what's more, I certainly don't earn my salary here.'

The *Marquês* raised a surprised eyebrow. None of his employees had ever said that to him before.

Susan's anger died down as suddenly as it

had been born. 'I'm sorry, *Senhor*, but only a little push is needed, I'm sure it would make Leonora a different girl.'

How appealing this girl was as she pleaded with him, thought the *Marquês*. Who would have guessed that this English Miss, with the speaking eyes and vunerable mouth, could stand up to him like this, but his own eyes were cool as he said, 'You will go and have a rest now, *senhorita*.' He placed a hand under her elbow and Susan was surprised and a little disturbed to find how extraordinarily aware she was at his touch.

He walked beside her through the wide oak doorway that had been a bastion against all invaders so long ago and Susan, for the first time, noticed the deeply bitten scars on the door's highly polished surface.

At the foot of the stairs he let her go. 'If you could help Leonora, Miss Deane, you'll have earned more than your salary and all my thanks,' and with a bow was gone.

# 3

The following morning Susan was lying in bed, enjoying the delicate perfume wafting in through the open windows, when a small, tousled head peeped around the interleading door and with a bound, Falina was on the bed.

'Marmu is bring the tea. I heard her coming up the stairs. Oh, I do love Saturdays!'

Susan switched on the bedside radio, liking to hear the news, not that she could understand very much, but felt it helped her Portuguese. She was only half listening when an announcement caught her attention. 'Ssh!' holding up a silencing finger. 'Now what was that all about, I wonder?'

The little girl wrinkled her small nose. 'Something about the hospital in Marula wanting people to give blood, I think,' she replied uncertainly her eyes big. 'Why should they want that?'

Susan explained as she drank her tea. Another announcement came over again at the end of the news. It was an urgent appeal to donors in the B positive blood group. A young woman who had just given birth, was dying.

'That's me,' said Susan quickly, dressed and went to find the senhora, Falina at her heels. She asked permission to go into the town and gave her reason why.

'May I go too, Tia?' asked Falina. 'I'll be very good.'

The old senhora shrugged indifferently. 'It is for you to say, Miss Deane.'

Taking this as a tacit consent, Susan ran up quickly to fetch her hat and keys and then down the steps to where her car was parked, Falina already in the car. Susan had never covered the five miles into Marula so quickly as on that Saturday morning and was hot and dusty when she presented herself to a Sister, who told her that it would have to be a direct tranfusion. She stared at the woman in dismay. That would take some time and she and Falina would be late for lunch. What would the Marquês say? Nevertheless she nodded. Lying looking up at the ceiling, she willed herself to relax. The old senhora would tell him where they were.

It was extremely hot in the small ward and the perspiration soon beaded her face. She dared not move either, in case it disturbed the flow of blood, but at last, after what had seemed an eternity, it was over and the puncture in her arm covered with a dressing.

'Now come and have some tea and a biscuit,' suggested the plump, pleasant faced Sister. 'Our patient is already much better.'

Susan looked down at the young woman, who still had a dreadful colour, she thought compassionately, then followed the Sister along a passage into a small waiting-room where she was given tea and dainty biscuits which were most acceptable. Falina joined her, filled with her morning's doings.

' — and I had lunch here and I've seen all the children and I'm going to be a nurse when I grow up,' she finished breathlessly, her dark eyes as round as saucers.

Susan smiled across at the excited child and wondered, amusedly, what her brother would say about that.

When they arrived back at the *quinta*, Susan dropped Falina at the front door with firm instructions to go straight upstairs and have a rest.

'I'll just park the car and then come too,' she said in answer to the enquiring glance, but by the time she had reached the hall, the *Marquês* was already there, coldly demanding why she had missed lunch and accused her of being a bad example to his sister. Before she could defend herself, he continued, 'Punctuality is rare these days, more's the pity Miss

Deane, but I'll not have it neglected in my house.'

He never shouted, but his cold voice was much, much worse, decided Susan, longing to get to her room. A tearing, ragged pain had begun to blind her and only vaguely heard what the *Marquês* was saying, answering 'yes, *senhor*,' and 'no, *senhor*,' at what she hoped was the appropriate moments.

He suddenly caught her by the shoulders, his anger taking a new turn. 'So, it was you who donated blood this morning,' his glance on the strip of plaster in the hollow of her arm. 'I heard as I came through Marula that someone had come forward. Did it have to be you, Miss Deane?'

Why couldn't this man have been out today and then she would have been spared all this, she thought, standing still between his hands, her eyes dark pools of fatigue, pushing wearily at the fair hair that lay wet on her throbbing forehead. There was no understanding this autocrat and she was too tired to try.

'It was the second appeal, *senhor*. I could not ignore it. I'm in an unusual blood group and this woman was dying.'

'Surely there were others?' he insisted. 'You have not yet recovered from your last heatstroke and now this? Are you quite mad, I wonder?' His features were a mask of displeasure. 'It was extremely foolish of you and why didn't that fool of a doctor keep you there until it was cooler? Sending you home in the heat of the day like this?'

'The doctor was not there and the Sister didn't know me.'

'It's just as well,' the *Marquês* said savagely, 'or else I'd have had his head. You were in no state to give blood.'

If only he would dismiss her, she thought raggedly. Surely he could see all she wanted to do was to be left alone? In a daze of pain she suddenly felt herself drawn closer to him.

'Come, *pequena*. If I'm not mistaken, that head of yours is rebelling again.' He propelled her unresisting up the stairs. 'You should have left a message,' he added quietly.

'But *senhor*, I did. *Senhora* Ferreira knew I had gone to the hospital. I asked for permission.'

'*Tia* only said you'd gone into Marula.'

How damning were half truths. No wonder he had been so angry, thought Susan. 'Perhaps the *Senhora* did not understand,' was all she could find to say.

They had reached the door of her room and with her fingers firmly on the door knob, she turned to the *Marquês*. 'I'll be all right and I'm sorry for the misunderstanding, *senhor*,' but he took no notice of her implied

53

dismissal and followed her into the room.

'Where are the tablets the doctor gave you?' and went across to the bathroom when Susan told him where they were.

She thankfully sank into a chair and kicked off her sandals. By this time the pain had increased considerably. It was so silly, for she had never had this reaction when she had given blood previously.

'The bottle is empty,' said the *Marquês*. 'I'll get you a repeat. Get into bed while I 'phone.'

Susan slipped hastily out of her frock and into a flowered housecoat, gingerly lowering herself on to the bed and buried her face into the soft pillow, as if to smother the pain that was engulfing her, weak tears forcing their way through her closed eyelids. She could hear the *Marquês* on the 'phone, wondering uneasily who it was having to listen to that tirade. Even getting a repeat didn't appear to be an easy matter.

'The Sister, unfortunately, can do nothing until the doctor returns,' he said as he came back and sat by her bed. '*Senhorita*, my father suffered much with his head. Will you let me try my mother's remedy? I don't know when Doctor Martenis will be back. He's out on a difficult maternity case fifteen miles away.'

Even in the midst of her pain, Susan was surprised at the softened cadence of his voice, vaguely conscious that all anger had left him. She nodded and then felt his long fingers probing the nerve centres at the back of her neck and shoulders, gentle rhythmic movements up her head and down into her shoulders.

'You have to help, you know,' the quiet voice went on and she could well imagine the amusement in those dark eyes, for she had stiffened at his touch. 'Release that tight band across your eyes.'

The words must have had a mesmeric charm of their own, for Susan found herself slowly responding, the tension leaving her body and miraculously, the pain was receding, as those hypnotic fingers worked their will.

How annoying this man was in his anger and yet surprising in his gentleness. She wondered drowsily, what kind of woman, if any, could make him happy, before sleep finally blanketed her mind.

The *Marquês* rose from his sitting position, dark eyes gazing down at the sleeping girl, relaxed at last, her fair hair a halo on the pillow, tears still glistening on her lashes, which he noticed were long and thick, making two innocent crescents on her lightly tanned cheeks.

Unusual colouring she had with those deep blue eyes and dark brows, allied to that fair skin and hair; remembered the angry gleam as she fearlessly stood her ground in the matter of Leonora. A crusader, if there ever was one, he thought and with an unreasoning sense of compunction, drew the light cover over her still body and closed the curtains across the window.

★ ★ ★

A tap at the door hours later, awakened Susan. '*Entrar*,' she called out sleepily. Her headache had quite gone and felt surprisingly refreshed and eager to be up.

It was Doctor Martenis and the old *senhora*, who was a trifle forbidding, her thin lips drawn up primly.

'*Tia*, I would dearly love a cup of coffee and some of those biscuits you are famous for,' and shooed her out of the room before turning to Susan.

He sat heavily on the bed, thanking her for what she had done adding that the mother was very much better, but he looked tired and dejected. 'I see you are better too, *senhorita*. A thousand apologies from Sister for sending you home when she did,' then added with deep feeling. 'How that man has raved and I had nothing to do with it. *Meu deus!*' Both hands came in contact with his knees. 'It nearly cost me my job for I'll have you know that the *Marquês* owns the hospital.' His round face was comical with dismay. 'But now I can take down a good report and all will be well again. I hope!'

'I'd like to get up,' said Susan, smiling into the eyes that had begun to twinkle again under their shaggy, untidy brows.

'No, no! You are to stay where you are. Time enough tomorrow to think of that.' He rose to go, but paused at the door. 'What did you take for the head?' he asked curiously.

Susan's shrug was evasive. 'Sleep was all that was required, thank you,' she said lightly. Nothing would induce her to tell him what the *Marquês* had done for her with those gentle fingers. It was something she would always remember, but barely acknowledging even to herself what emotions his touch had triggered off.

Later that evening she sat amongst her many pillows reading, her supper tray at the foot of the bed, Falina splashing in the bath next door, the maid scolding for all the water on the floor. A soft tap at the door made her glance up. '*Entrar*,' she called out, thinking it was Leonora, but, but it was the *Marquês*.

'I have been told that you're very much better, *senhorita*.'

There in the doorway he stood, immaculate in formal dress, the dark head held high.

Susan thanked him. 'There must be magic in your touch,' she said with a shy smile.

A disconcerting glimmer of amusement lit his face as he glanced down at her. 'I'm glad to see you are more relaxed,' he observed blandly, 'but you're still too pale. I was just beginning to notice the return of colour to your cheeks and now this must happen.'

'I'm tough,' she replied with a disarming grin.

His eyebrows took on the now familiar slant. 'The good doctor is quite amazed at your quick recovery. I noticed that you did not tell him of my magic hands.' The glimmer had now become an open smile and the warm treacherous colour flooded her face, as she gathered the top of her bedjacket closer to her throat with a nervous movement, resenting the shyness that had suddenly filled her, but she was spared an answer for the *Marquês* went on, 'I'm on my way to dine with the Gonzales family, but I first needed to satisfy myself that all was well with you.'

'Thank you. Is Leonora going with you?' her voice eager.

'The invitation did include my sister, but

I'm afraid, once more, I must carry her apologies.'

A spurt of anger flared inside Susan at his complete indifference. '*Senhor*, could you not have made her go? I'm sure she would enjoy herself, once she arrived there.'

'And be accused of being cruel?' he asked swiftly. 'No thank you, Miss Deane! I have no desire to be accompanied by anyone who has no wish to dine with two, very dear old people.'

So the Portuguese, even with their strict upbringing, have their problem teenagers. Her dismay must have been mirrored on her face for he said more quietly, 'Now you are not to worry about my sister. Sleep well, *senhorita*.' He walked to the door, bowed slightly and was gone, leaving Susan with the feeling that she had been dumped back a hundred years.

She had no intention of staying in bed, throwing back the light covering, slipping into her dressing-gown she padded into Falina's room.

The little girl threw herself into her arms, Susan gathering her close. 'Raynor said I wasn't to bother you, but I'm glad you've come to say goodnight,' nuzzling her face into Susan's neck, like a small pup, her eyelids already almost too heavy to keep open.

Susan slipped Falina between the sheets, smoothing the straight, dark hair away from the small round face. 'Sleep well, darling. See you in the morning,' and tucked the mosquito net in under the mattress. Turning off the light, Susan made her way to Leonora's room.

She uncurled herself from the couch, turned down the transistor and invited Susan to make herself comfortable, saying how glad she was to see her so much better.

A silence fell, Leonora tapping out the music with her slippered foot.

It was left to Susan to break that silence which she did a little warily. 'I hope you don't mind, but I spoke to Doctor Martenis about you the other day.'

'Did — did you? What did — did he have to — to say?' There wasn't even a flicker of interest in the stammered question.

'Only that there is nothing physically wrong with you. Really Leonora! Why do you lie around like this all day? Why aren't you up and about enjoying life? You, who have so much,' waving a hand around the beautifully appointed room with its white wall-to-wall carpet and heavy blue satin drapes at the windows.

'What is — is there to do?' she asked bitterly. 'Nothing ever happens here at — at the *quinta*. What is there to — to do but

60

listen to the radio and read?'

'You should be spanked,' replied Susan with exasperation. 'You could have gone out with your brother tonight, for a start.'

The girl shrugged and half turned away, the heavy hair concealing her face.

'Well, Leonora?' Susan asked gently, a pang of pity going out to her as she noticed the full lips tighten sullenly. Susan tried another tack. 'I know I'm harping on the subject, but I really want to help. I think I know why you shut yourself up here, but it's not going to help, you know.'

Leonora gathered up the dark hair from off her face. 'Do you? I — I don't think so.'

'You're afraid to meet people.'

Leonora turned on her then, leaning forward, the dark eyes flashing in the creamy pallor of her skin.

'And what else do you think I should do? You don't know what it is to stammer, not to be able to get out the words quick enough, watching the other person slowly become more and more embarrassed, wondering if they should help me or not. 'Leonora Monteiro stammering, she never did that before' and behind my back they cross themselves. Oh yes, they do,' she panted in reply to Susan's swift protest of denial. 'Falina told me.' She turned into the cushions, burrowing her face in their satiny smooth-ness, sobbing hysterically. 'I'll — I'll never get a *marido* now.'

Susan gazed at the huddled figure with compassion, but knew this was no time for spoken sympathy. 'How can you even expect your brother to make any arrangements when you are so sickly?'

Leonora's reaction was as sharp as Susan had hoped it would be.

'I'm not sickly,' she blazed. 'It's only my stammer. Who wants a wife who cannot bring out the words of love quickly enough?' she wailed.

Susan did not reply for she was studying the Portuguese girl with almost clinical interest. 'Do you realize, that you have only stammered once since you've become so worked up?' she remarked with suppressed excitement.

'No, I — I — ' The stammer was back, the words coming out with painful slowness.

'Now tell me what made you so afraid.'

'Yes, yes I — I was afraid.' Her breath came in hurried gulps. 'Afraid I'd never walk again. Afraid that — that no man would ever want me. My leg — it was taking such a long time to — to heal and then when it did I couldn't stop this stammer.' She swallowed convul-sively. 'Oh, Susan, it's been awful!' clutching at Susan's hand that had been held out in silent affection. 'How is it you understand? My brother, *Tia* — ' she shrugged. 'Not one of — of them has guessed what I've suffered, here — here in my inner self,' placing both hands at the base of her throat with a touching gesture that wrung Susan's heart. 'Doctor Martenis comes and says my leg is healed. I could have told him that,' she added scornfully.

'Sometimes a newcomer sees things from a different angle. Anyway, you were a silly, weren't you? Tomorrow we'll start singing lessons. I do know that correct breathing helps in some cases. Will you give it a try, Leonora?'

The girl stared up at her and Susan was reminded of the eyes of the old spaniel they had had at home years ago, great limpid eyes of hope when a walk had been suggested. Leonora looked rather like that now as she sat in a kind of daze.

Oh, I'll — I'll do anything! Do you think it — it will help me? There's a piano in our sittingroom. Could we start now, this — this very minute?' Leonora gave an excited laugh as she rose hurriedly from the couch, pulling Susan up with her. 'Perhaps I'll — I'll soon have a *noivo*,' she added buoyantly.

The two girls eagerly opened the old

instrument, Susan warning that it might take months before there was any real improvement, but Leonora waved this aside, listening intently to Susan's instructions.

It was the slamming of a car door, hours later, that made them both jump guiltily.

'You — you have been ill and should have been in bed long ago,' said Leonora remorsefully. With a giggled 'goodnight' they both ran lightly back to their rooms.

The following morning Susan felt as if she had never been ill, quickly showering and slipping into a crisp mauve frock with embroidered pockets. Brushed her hair vigorously and ran a lipstick over her mouth.

'It is just as well you are recovered, Miss Deane,' *Senhora* Ferreira said sourly at the breakfast table. 'I want you to go to Marula for me this afternoon.'

'Coming, Leonora?' asked Susan.

She shook her head with a conspiratorial air. 'I'm going to practice all day.'

Susan had wondered if the enthusiasm of the previous evening would have waned. If it had done nothing else, it had at least, brought Leonora down for breakfast.

Susan had just finished her coffee when Luis, the butler, came in to say she was wanted in the library. She excused herself and found the *Marquês* had two middle aged people with him.

'*Muito abrigada, senhorita,*' murmured the woman brokenly and came forward to take Susan's hand, who looked enquiringly up at the *Marquês*.

'It was these people's daughter you saved,' he said with a smile.

'How is she?' Susan asked eagerly.

'Much, much better and she wishes to know if the little one might be called after you?' This was translated by the *Marquês*.

'I'll be honoured,' replied Susan taking the woman's gesticulating hand in hers.

'Miss Deane, you now have a namesake here in Mozambique,' the *Marquês* told her.

★　★　★

'I'd like to buy my namesake a present,' Susan said to Falina that afternoon on their way to Marula. The little girl was delighted at the prospect and they spent a happy half hour looking around the shops.

It was very hot in the small town, the sun beating down fiercely on to the pavements and they were glad to sink into a chair under a brightly striped umbrella, all their shopping completed.

The waiter had just brought their order

when a man approached their table. Broad-shouldered and tall, he seemed to tower over them, casting a shadow over the pavement.

'Miss Deane?'

Susan nodded, her glance taking in the tanned face, with its nondescript blue eyes, seamed at the corners by too much sun and sandy hair.

'Good, may I join you?' pulling out a chair and throwing his dusty, khaki hat on to the pavement, before she had time to say anything. 'I worked with your brother. Mark Fenn is the name.'

'I — but how did you know I was here?' her surprise showing in her wide, friendly smile.

'Bush telegraph. Never known it to fail, especially when it concerns a pretty *senhorita* with fair hair and an upperty nose. You just don't know what it does to a fellow to see an English face again,' he added fervently, with an admiring glance at her laughing eyes.

'Did the news of my arrival come from the *Marquês*? I just can't believe that the bush telegraph works that fast. Your camp is miles from anywhere, isn't it?'

'As a matter of fact, he did say that you would like to have a talk to someone who had known Peter. We do have a phone in that god-forsaken place.' His voice sobered. 'It was a nasty shock to us all when we heard the news.'

'Did you know why he was killed, Mr. Fenn?'

'Mark, please!' he protested. 'Two Britishers in a strange country and all that. The authorities just notified us that he had been killed in a street fight.'

Susan winced. How many times she had heard that phrase. 'So you can't throw any more light on the subject either,' with keen disappointment, 'but I'm sure that it is not the whole story. Peter was not the type to become involved in a street fight. I want to know the reason behind it.'

Mark shook his head. 'Tempers ride high here with so much wine flowing,' he reminded her.

'Will you make enquiries for me in the camp?' eagerly.

'Sure I'll do that. You're hoping someone might have been on the spot who could, perhaps throw some light on the subject? Don't bank on it, Susan,' he warned. 'If there had been any monkey business, the authorities would have been informed.'

Falina had finished her cool drink and was listening with wide eyed puzzlement. Susan gathered up her belongings reluctantly.

'We must be going,' with a meaningful glance at the little girl. 'It's been nice meeting you, Mark. Perhaps we could talk some other time?'

'Hey, don't go yet,' he protested, 'and if you must, what about a date?' pushing back his chair. 'Is a fellow allowed to call on you at that palatial pile?'

Doubt made her say, 'I don't know.'

'But surely you have some free time? Go to the Union, my girl, if you haven't.'

Susan chuckled. 'I do so little as it is I couldn't ask for time off.'

'You jolly well fight for your rights and insist on an afternoon and evening off at least once a week. Let me take you dancing,' then seeing the doubt still in her eyes added, 'I'm quite respectable, you know.'

Susan laughingly accepted the invitation and a date was arranged for that evening as Mark had to return to the camp.

She would ask the maid to listen out for Falina and felt a thrill of excitement rise within her. How nice it would be to go out again with such a presentable male. It did something for a girl, she decided as they drove back to the *quinta*.

★  ★  ★

68

Leonora was frankly jealous when Susan told her about her date. 'Why can't I have a date too,' she pouted. 'It's not fair for you to go out like — like this and for me to stay at home always. I — I wish I'd been born English,' hunching a petulant shoulder. 'I — I hate you!'

Yes, this is what Leonora needed. A boy-friend. How thrilled she had been when Mark asked her to go out with him. A man's interest in her and her alone and wondered how it could possibly be arranged, for Portuguese girls were not allowed to date indiscriminately.

Both girls had watched her dress, Falina refusing to be banished to bed, telling Susan that she looked beautiful.

Clothes really did make a difference, decided Susan taking a last glance in the mirror, or was it happiness that did more? The black sprigged blue dress that had been a gift from Nannette, brought out the sparkle in her eyes. Her fair hair lay smoothly to her shoulders and a brighter lipstick completed the picture.

A car could be heard as it pulled up at the front door and she hurriedly picked up her white cardigan.

'No!' shrieked Leonora, shuddering theatrically. 'How can you possibly wear that to a dance?'

She rushed from the room and returned with a lovely black lace stole, draping it becomingly over Susan's shoulders, then stood back to admire the effect.

'That's much — much better.'

'Oh, Leonora, it's lovely, thank you.'

She shrugged indifferently. 'You may have it. I have others I — I never wear.' Her spirit of enthusiasm had already evaporated.

'I must go,' said Susan quickly. 'Falina, you must go to bed immediately,' kissing the little girl and holding her close for a moment. 'Leonora, will you come down and meet Mark?'

'No thank you. I might hate you even more then,' her glance studiously turned away and as usual, when she was in an angry state her stammer lessened considerably.

Susan looked at her with exasperation. There were times when she longed to shake Leonora.

Mark was in the hall talking to the *Marquês*. They turned as she descended, her full skirts swaying around her ankles at every step.

'Hi, there!' murmured Mark appreciatively. 'So I didn't just dream you up. Aren't I in luck's way?'

Susan thanked him with a smile then turned to the *Marquês*. 'I hope you do not mind, *senhor*, if I go out tonight? I did ask the *senhora* if I might do so.'

He inclined his dark head. 'No, I do not mind, *senhorita*. Enjoy yourself, but please,' turning to Mark, 'bring her back early.'

His goodnight was cool and impersonal as he stood on the top of the terrace steps, raising a hand as they drove down the avenue.

He was very much the *fidalgo* of the *quinta*, she thought with a backward glance at the tall figure outlined now in the open doorway, the light spilling around him.

★　★　★

Susan enjoyed the evening. The food was good and so was the band. In between dances, she learnt a lot about Mark Fenn, who was quite happy talking about all the places he had been to. He appeared to have no close family ties and building bridges and dams was his first love, the bigger, the tougher the project, the better he liked it.

'But the loneliness,' protested Susan, thinking of her brother.

'I've got used to it, but you know, little one,' he said quietly, his fingers turning the fragile glass making the red wine sparkle like rubies, 'seeing you sitting there, makes me wonder if I wouldn't like to settle down, with someone like you.'

She scoffed at him gently, refusing to take him seriously. 'Not you, Mark! When this dam and bridge is finished and you go back home, you'll date all the girls you can and before you know it, you'll be looking for your next contract, in some inaccessible spot on the globe.'

He grinned at her, laying his hand on hers as it lay on the table. 'You're sweet, Susan. May I see you again?' His pale blue eyes met hers steadily.

'I'm usually in Marula on a Saturday morning and you're to let me know if you find out anything about Peter, remember? Mark, I don't suppose you were able to ring the camp?'

He nodded. 'But I told you how it would be, honey. No dice. Most of the men were in town that night, but nobody saw the fight. Please don't look so disappointed.' A silence fell as Mark took a long draw on his cigarette. 'By the way,' he asked casually, 'How's Nannette?'

Susan shot him a startled glance. 'You knew her?'

'Very well. I was with Peter in Brazil. How she hated the life out there. I suppose she's remarried?' The words were too casual and when she shook her head he went on, 'Oh

72

well, I don't suppose she'd give me a second look, anyway.'

'You liked her a lot?'

'I fell for her, boots and all,' he replied fervently. 'Don't get me wrong,' he added with a rueful laugh. 'I worshipped from afar, but what a girl!'

It was not surprising that Mark had fallen for her sister-in-law. He probably hadn't seen an English woman for months and when someone like Nannette arrived on the scene, was like water to a parched desert.

He chuckled. 'Of course the very first thing she demanded was a hot bath. I ask you, in the middle of a jungle! The rest of us were quite happy to take our daily dip in the river. I must admit though, there were crocs and her ladyship said they might cramp her style.' He lit another cigarette from the tip of his previous one.

'I'm very sure she got her way. She always does.'

'We rigged up a forty-four gallon drum and stoked a fire underneath it,' then with a gleam of humour, 'Of course, no bath was to found in camp because the cook, a local johnnie, was cooking the stew in it. He'd never seen a zinc bath before.'

Susan laughed. 'What on earth did you do? Poor Nannette, no wonder she hated Brazil, or I should say the jungle. No mod cons.'

'Another drum cut in half lengthways sufficed. Lady Nannette!'

'She still speaks with loathing about that trip.'

Mark nodded. 'She hated it pretty badly, but I'll say this for her she would dress for dinner each evening and what is more, expected the men to do the same, even if it was only a clean shirt and pants. A lot of the chaps were glad when she left.'

'I believe she went and lived in the nearest town.'

'Yes, but Pete couldn't see her all that often. It was a three day canoe trip, but what really finished Nannette was the way the washing was done. Everything was taken down to the river and pounded on a smooth stone.' He shook his head, an impish grin widening his mouth. 'You can imagine what happened to her dainty frillies and no shop around the corner, either. That's when things really blew up.'

He drew hard on his cigarette, curling his tongue around the smoke before exhaling it. 'Any chance for me, there?'

She shook her head in doubt. 'Mark, you'd have to give up your present way of life, to begin with,' she added gently seeing his downcast face.

'I'd even do that,' he conceded soberly. 'Hey, let's dance again. We're getting far too serious, but you are a very understanding type of person. One more question. Would you be willing to go where your husband led?'

A warm smile lit her face and she nodded. The band had struck up a lively peasant tune and he pulled her to her feet, but Susan had seen the ornate clock above the large mirror on the far side of the room.

'It's after twelve,' she gasped. 'We must be going!'

'Who do you think you are? Goldilocks?' he demanded.

'You've mixed up your characters. We must leave now,' she insisted.

Grumbling he placed her shawl around her shoulders with a hug. 'But the night is still young,' not willing to relinquish so genial a companion.

'The *Marquês* said I was to be in early. Please, Mark.' There was an urgent note in her voice now.

'Oh, all right, if you must. Don't look so harrassed. If his lordship thinks this is late, he must be absolutely archaic.'

Vila Media was at least forty kilometres from the *quinta* on a road that was shocking. The *Marquês* was going to be furious, glancing at her watch for the hundredth time.

'He at least won't eat you,' Mark murmured with a touch of humour, the car shying over the corrugations. 'What's he like to work for, Susan?'

'I don't see him all that much,' her reply noncommittal. She had no wish to discuss her employer.

'Well, he has no right to act the heavy father.'

'He's not that old,' said Susan hotly.

'Hey!' He shot her a quick glance, but her face was only a blurr in the lights from the dashboard. 'Now don't go losing your heart in that direction.'

'I've no intention of doing so,' wondering why she should suddenly feel so ruffled.

'Good! It's been rumoured that he is, at last, thinking of marriage. A *Senhora* Oliveira and her daughter, Pepita, were staying in this vicinity a couple of weeks back. I believe they are well-known to the members of the Monteiro family who live in Portugal.'

A silence had crept into the car. 'Why can't you say something?' asked Mark accusingly, changing down quickly as a particularly large rut in the road loomed in their headlights. 'Here I'm giving you a juicy titbit and you go quiet on me.'

'Sorry, Mark,' her laugh shaky, gathering her stole more closely around her shoulders as if the night air had gone several degrees cooler. 'I suppose it is only natural the *Marquês* should marry. Have the mother and daughter gone back to Portugal?'

'No, they are visiting friends in Lourenco Marques. I suppose to give the girl a chance to see if she likes this country and also to give the *Marquês* time to make up his mind.'

'It — it's all so cold blooded! For the girl too. Think of all the talk if nothing comes of the visit.'

'These kind of arrangements are understood here and carried out most discreetly, but you should have seen her, Susan,' suddenly waxing lyrical. 'A smouldering Latin beauty, with a figure and temperament to match.'

It was nearly one o'clock by the time Mark turned into the long avenue of flame trees and Susan braced herself to meet the *Marquês*, knowing full well that he would be waiting up until she came in.

# 4

Susan said a hurried goodnight to Mark, refusing to let him see her to the door. Mark allowed this because he just did not believe that she would have any trouble with her employer.

Susan ran quickly up the shallow, floodlit steps. The *Marquês* was in the hall.

'So, it is just as well that young man has gone,' his glance going beyond her. 'Do you realise the time, Miss Deane?' he enquired coldly, his scowl clearly indicating his disapproval.

Her gaze showed something of the strain she had been under this last hour and could feel the nerve ends throbbing in her temples, but her reply was calm enough. 'Yes, *senhor*, but after all, it isn't so very late.'

'So, you think one o'clock not late? This laxness between the sexes may be permissable in your country, but it is not tolerated here.' The words were frosty. 'I've been expecting you since twelve o'clock and if it had not been for my guest, I'd have gone to look for you.'

Susan stared at him with surprised dismay.

78

This was too much. 'But *senhor*,' anger splashing colour into her cheeks, 'I was out with one of my countrymen. Our ways are not your ways, as you have just said. I can look after myself,' and despised herself for being so much on the defensive.

The *Marquês* turned to lock the massive oak door. 'It's late, we'll discuss that another day,' he replied coldly, turning to her again. 'You are amused, Miss Deane?'

A grin had unexpectedly lifted the corners of her mouth.

'I said to Mark on the way home that I should have remembered to ask for a key, but I would not have liked to have carried that huge one around.'

'We do not give keys to our girls,' came the repressive reply, but added politely, 'Did you enjoy your evening?'

'Thank you, yes,' and was as formal as he had been and wishing him goodnight she turned to go upstairs, but as she reached the second step, the *Marquês* spoke again, the coldness gone from his manner, the light from the wall bracket etching shadows on his lean, upturned face.

'Was Fenn able to give you any news of Peter?'

Susan shook her head and her voice lost some of its buoyancy.

'Do not worry about it, *pequena*. Even if you have to go back having learnt nothing, it will not, I'm sure, destroy that faith and loyalty you have for your brother. I find it most touching. Goodnight.'

As she walked up the stairs, Susan wondered at the almost irrational desire that filled her down in the hall, to ask the *Marquês* about Pepita Oliveira. Was he going to marry her? Her lips twitched with amusement and knew that she wouldn't have dared to do such a thing.

★ ★ ★

The following morning Susan found Leonora sprawled on her bed, the record player beside her, listening to the famous *fado* singer, Ameria Melo. Susan did not appreciate this type of music; it was far too sad.

Leonora took no notice of her and Susan was filled with exasperation. 'Don't you want to hear about my evening?' shouting the words above the beat of the music and when Leonora did not respond, switched off the machine.

'I'm — I'm not interested in what you do — did — not at all.' The awful stammer was back.

So Leonora still resented her having gone

80

out on a date. The girl's dark, unhappy eyes pulled at her heartstrings. 'But I'm very interested in what happens to you, Leonora,' she said gently.

'You — you were hired to look after Falina, not me,' raising herself quickly to turn on the record player again, but Susan stretched out her hand.

'I was hired to be a companion to you both. Come on, get up. I've a fancy to see the horses this morning,' coaxed Susan. 'Old Luiz has been telling me about the wonderful stables you have here.'

The old butler, grizzled of hair and a stoop proclaiming his age, had told her he had been with the family for many years, brought out by the present *Marquês'* father to work in the stables, but later was trained as a butler. Susan often chattered to Luiz, for his work was light and most days at certain times, could be found dozing in a comfortable chair on the side patio in a shady corner, content with his memories.

Leonora refused to move. 'And what if I say no?' There was a glimmer of provocative interest in the upward sweep of her Latin eyes.

'If I think I can't be of some help to you, then I'll hand in my notice tomorrow.'

'But — but why?' she asked in bewilderment.

'Oh, for heaven's sake, Leonora, be your age. I do little enough as it is.'

'Is that how you feel when you have to work for your living? I don't think that would be at all comfortable.'

'Do come. I don't know a thing about horses and I'd like to learn.' Some of her eagerness must have rubbed off on to the girl, for she said more animatedly.

'I used to ride a lot,' then dropped back into her old sullenness. 'I — I don't any more.'

Susan prompted her gently. 'Why not? Can you explain to me?'

'I had a nasty fall, as you already know, going over one of the jumps and broke my leg. I had been hoping so much, to compete in the horse trials. I — I don't know how I came to — to do such a silly thing. I've ridden for years.'

'I should imagine it's easy to fall off.'

Leonora's glance was scathing, her tone pitying. 'Now I know you — you don't know anything about riding or horses,' but Susan was glad to see that she swung her long legs off the bed and after a half-hearted attempt to brush her long hair, led the way downstairs calling to Falina as they did so.

They went from building to building inspecting each animal, José, a wizened little man with bulging legs, giving Susan the case history of each one. Leonora had begun the tour in complete silence, but in some magic way the old groom finally broke through her indifference, until her enthusiasm equalled that of his, especially when they watched two thoroughbreds thunder down the course.

'*Senhorita* Leonora, it is good to have you here again,' José said humbly, 'and now I must show you the king of them all. Come.' Pride edged his words.

They walked between the orderly white-washed buildings, each door rivalling its neighbour in colour and came, at last, to the end box. José leant on the half-door and the beautiful animal tucked its nose into the crook of his arm. There was a perfect understanding between man and beast. He spoke in an awed tone.

'The finest stallion I have ever seen and what is more, bred right here in our own stables. The *fidalgo* is very pleased.'

'Does — does my brother intend to sell him, do — do you know?' Leonora asked, stroking the satiny nose. 'Oh, I hope not. I — I shall not let him,' with a little of the *Marquês'* arrogance.

Susan marvelled at the change in Leonora

as she went knowledgeably, over each point of the animal with the groom, but allowed herself to be dragged away by Falina to another section.

'This is my very own pony,' she said proudly, patting the neck of a small, shaggy animal as round as a barrel.

'You don't ride him enough,' smiled Susan. 'He's much too fat.'

★　★　★

During the next few weeks saw a marked improvement in Leonora, so much so, that even her brother remarked on it. The stammer was still there, but her words were controlled to a certain degree. What the girl now needed was confidence in herself, not as a child, but as a woman and Susan knew that she would have to approach the *Marquês* again. The prospect was alarming and decided that a great deal of tact would be necessary. The following morning, as they were finishing breakfast, she asked if she might see him.

The *Marquês* shot a speculative glance at her serious face. 'Very well, Miss Deane. Come into the study now,' but instead, took her through the room out on to the patio and seated her in one of the cane chairs, before taking one opposite her.

'Now, what is it you wish to speak about that makes you so apprehensive?'

Susan had clasped her hands tightly in her lap, his smooth voice, this time had not robbed her of her confidence. There was too much at stake.

'It's about Leonora.'

'Oh, so you're not thinking of leaving us yet? I quite thought you had had enough of the *quinta*,' and our outdated ideas, that mocking gleam seemed to imply.

'There is such a big improvement in your sister, *senhor*, I think you have seen it yourself, but there is still a barrier holding her back.'

'Go on, Miss Deane, or is your idea so outlandish that you hesitate to tell me about it?'

With a mulish stubbornness and eyes that gazed straight at him, she said abruptly. 'Leonora needs a love interest.'

So much for tact, she thought wryly. The words had even surprised her and waited for the explosion, but strangely enough, none came. Instead the *Marquês'* mouth twitched.

'Don't we all?' he retorted pointedly and she knew he was thinking of Mark Fenn.

'It's perfectly true, *senhor*,' a faint colour

creeping into her cheeks. 'I enjoyed that evening with Mark immensely and that proves my point.'

'You feel my sister needs a boyfriend, is that it?'

Susan nodded and the *Marquês'* tone was perceptively cooler. 'My dear Miss Deane, I'm afraid you have not been long enough with us to know our customs.'

'But suppressed teenagers break out in a rash of wild spirits when they get a little older, or get a little freedom,' she reminded him swiftly.

He inclined his head, a gesture that was so familiar. 'We're also taught, from an early age, that we owe much to our name of which we are justly proud.' He was being very much the *Marquês* now, aloof and feudal, even his accent had become more pronounced.

The sun had found the shady nook where they were sitting, lighting up the flowers surrounding the patio. Flamboyant and brazen, no colour here was muted. The people themselves, were a little like that, she reflected, banked fires just below the surface, ready to spring to violent life at the first opportunity.

Leonora herself, was a smouldering mass of contradictions. She was already a woman. Was the *Marquês* aware of this, or did he think he could keep her an obedient, though sulky child?

'Why can't your sister have friends of the opposite sex?' Susan demanded. 'I just don't understand this old-fashioned idea of cooping up young girls. It's asking for trouble. You are fortunate Leonora does not meet men behind your back, *senhor*, but shutting herself in her room, isn't the answer either.'

The *Marquês* replied patiently as if giving a lesson to a backward child. 'We've had these laws — rules,' he shrugged, 'for a long time, *senhorita*. The good name of our girls means a lot to us and you must admit, that we do not have a lot of disastrous marriages, so there must be some merit for our method. Even the British up to the beginning of this century and even later, guarded and arranged marriages for their daughters, so the aristocracy would not die out.'

'But surely, if she falls in love — ' began Susan hotly.

'Isn't it better to fall in love with the right man and I think propinquity plays a big part,' he countered, 'and who better equipped than the older members of the family to judge the type of man who would be best suited for their daughter, who would make her happy?'

Susan eyed him with dislike, but could not help asking curiously. 'Do you pair couples off who have similiar ideas? It is often said that opposite attract.'

'But do they always make good marriage partners? I very much doubt it. There must be a mutual basis for understanding, for surely, if the interests are too divergent, the marriage could be a failure, for one of the partners would always have to give in to make a harmonious life. Can two walk together except they be agreed?'

'As far as you are concerned, *senhor*, it would be the woman who always gave in.'

'That would never be your fate, Miss Deane,' he retaliated with lazy amusement. 'As for your suggestion, it is impossible. I realize,' he added hastily, seeing she was about to protest again, 'you have Leonora's welfare very much at heart and what is more, you are not a coward, my little English *menina*, but we cannot have you change our way of things,' his smile conciliatory, in complete variance to the firmness of his words.

'My English blood rises at the very thought of arranged marriages.'

The *Marquês* shifted in his chair, the sun gleaming on his well groomed head.

'You must see our point of view. Leonora will be a wealthy woman one day. What if this — ' waving an expressive hand, 'this boyfriend were to take advantage of the situation?' He shook his head. 'No, my sister will marry into an old, established family when she has recovered from her accident and will never have the worry that her husband might have married her for her fortune. That could happen.'

His mouth was so cynical that Susan could have hit him. 'It surely won't be for love then, *senhor*.'

'We don't push our girls into an alliance with someone they dislike.' He had drawn behind his wall of arrogance again. 'Meetings are discreetly arranged,' he assured her coolly.

Susan should have let the matter rest there, but there was more that needed to be said. 'Leonora tells me that you haven't made any marriage arrangements for her at all. Perhaps you don't know this, but she's taking it as a personal insult. This is making her unsure of herself.'

The dark eyes were unwavering. 'How could I when she made no attempt after the accident to live a normal life?'

'And that is what I can't stand about these arranged marriages, *senhor*. Leonora is not fit to be offered to the son of a noble house. How glad I am that I can marry where I will,' she said passionately, sitting upright in her chair, her eyes large with indignation, the words coming out unguarded.

The *Marquês* had risen and was bending over her, his long fingers gripping the back of her chair, his anger a coiled spring kept rigidly under control. She shrank back from those glittering eyes.

How foolish of her to have allowed her wretched tongue to run away with her like that.

'I will arrange Leonora's marriage when I see fit,' his voice dripping ice.

'I'm sorry, *senhor*, it is not my place to criticise,' suddenly contrite, 'but I don't think you understand. How could you, for you don't take the slightest interest in your sister at all.'

★ ★ ★

The double doors of the small sittingroom were wide open, letting in the fresh evening air, rippling the long, heavy satin curtains. Susan was trying to unravel Falina's jumbled up knitting, the little girl crouched at her feet. Leonora was playing a lively flamenco tune on the piano with more verve than accuracy.

'I do like to see my women happy,' came a suave voice from the doorway.

Three pairs of eyes turned to the *Marquês*, Falina scrambling to her feet to bring her

90

brother into the room. A sardonic gleam had sprung into his eyes as he glanced down at Susan. He had seen her stiffen, but all she said was 'good evening' and asked if he would like a cup of coffee. He refused, his gaze on her bent head, her shining hair a cap of gold under the lighted lamp, her fingers busy with the wool and needles again.

Susan wondered why he had joined them. It was unusual.

'*Senhor*, I've been told that the orange blossoms are a most beautiful sight. Is it possible to see them?'

'Miss Deane,' he retorted in a tone to match her own, a spark of fun lighting the austere face. 'You have permission to drive down the road any day and see them for yourself.'

'Oh, please may we go with you, Raynor?' cried Falina.

'I think it would be nice,' agreed Leonora unexpectedly, which made the *Marquês*' eyebrows rise in surprise.

'We'll have to see into this, then,' he said with a mocking smile at Susan. 'Perhaps you would all like to accompany me on Wednesday morning? We can't let the school holidays go by without some treat and I think Miss Deane should be rewarded.'

She did not mind being teased at all, her

smile eloquent, but was surprised when he issued an invitation to accompany him to Marula.

'And let's have coffee at the *pousada*,' cried Falina, eyes shining.

The *Marquês* put his arm around the excited little girl.

'Will we have to change, *senhor?*' asked Susan.

'No, it will be my pleasure to be your escort.'

Leonora came alive under his admiring glance. 'Give — give us five minutes and we'll meet you downstairs, Raynor. He's not done this for — for a long time,' she added to Susan as she brushed her hair.

'You've had invitations to dine out, Leonora.'

She shrugged, her dark eyes mischievous. 'I'd much rather go to the *pousada* for coffee.'

★ ★ ★

'Do you know what I like most about you, Miss Deane?' murmured the *Marquês*, holding open the car door.

She shook her head, the full moon washing her tilted face with its soft light and brought to his mind the creamy petal of the orange blossoms.

'The way a lamp is lit behind those blue eyes as though heaven had been offered to you. At such simple things too, like an unimportant visit to the *pousada* in Marula.'

'It's not unimportant, *senhor*. I'm so pleased you invited us and at Leonora's happy reaction.'

'And you, *senhorita*? Will you get enjoyment out of it too, or would you have preferred another escort?'

The name of Mark Fenn hung around them like wisps of cobweb. 'I'll enjoy it too,' she said simply and saw his taut smile disappear.

The proprietor of the *pousada* was most deferential in his greeting and led them upstairs on to the balcony. The night had blanketed it with coolness and it was a pleasure to watch the passing traffic. There was only one other person on the balcony. A young man, typically Portuguese and Susan noticed Leonora glance at him with considerable interest on several occasions. She turned to the *Marquês* to distract his attention, talking at random on different subjects, until his quizzical smile made her cheeks burn and she threw him a look of reproach. He had a disturbing knack of reading her mind.

'Perhaps he's admiring that fair head of yours,' he murmured.

'Oh no. It is Leonora who has his attention. I'm so glad.'

'Leonora's happiness means much to you, doesn't it? I wonder why?'

He had leant back, relaxing in his chair, but was regarding her closely.

She eyed him uncertainly. Did he think she was trying to worm her way into the life of the *quinta* and so become indispensible? Her prickles rose and was just about to retort back hotly when Falina came and stood at the balcony rail and the moment was lost.

'Look, Susan, there's a horse drawn carriage. When I get a *noivo*, I shall ride in a carriage.'

The *Marquês* threw up his hands but he said kindly. 'I'll make a note of your wish, *pequena*. Did Miss Deane give you permission to call her Susan?'

Falina nodded. 'Susan says Miss Deane sounds too old maidish.'

'But I'm of the opinion that our Miss Deane will never be an old maid.'

The *Marquês* was quite approachable, thought Susan with surprise and as for Leonora, smiling at this sally, was positively glowing.

They did not stay long at the *pousada* and when the car stopped outside the house, Susan hurried a sleepy Falina to bed, leaving the brother and sister in the *sala*.

What a different person Leonora had been this evening, thought Susan safe under the mosquito netting and hoped the *Marquês* had also noted the fact.

★ ★ ★

The following morning Susan called in at the hospital again. She had become quite friendly with the doctor, but he was more than a little apprehensive when she told him what was on her mind.

'Do you mean to say you actually asked the *Marquês* if Leonora could have a boy-friend?'

She nodded.

His mouth dropped. '*Caramba!* *Senhorita*, you are of the most courageous!' he added admiringly.

'You make it sound as if I have an awful lot of cheek, but that isn't so. Leonora is much better, but there is something holding her back. It's a pity that a marriage couldn't be arranged. It would make a big difference.'

'And hasn't the *Marquês* made any arrangements?' The little doctor gazed at her, fascinated, incredulous that a mere employee should have taken so much on herself, for he knew the *fidalgo's* temper. It could not have been easy for this girl and watched her shake her head, with added respect.

'He's waiting until Leonora has quite recovered, but the fact that nothing has been done in this matter, worries Leonora.'

'Our women hang much on marriage,' he conceded. 'It obtains for them a status, you understand, but I — I think you should also know that Leonora blames herself, a little, for her mother's death.' He held up a hand. 'The *Senhora* Monteiro, Raynor's mother, was advised by her medical adviser, that another child would be dangerous, but — ' he shrugged, 'things happen and the *senhora* died giving birth to Leonora. Unfortunately, *Tia* Maria told Leonora, which has not helped matters.'

'Oh, poor Leonora! Falina then is her step-sister.'

The doctor nodded, 'but I cannot produce a boyfriend for her, like a — a rabbit from a hat. The *Marquês* would have my neck. No, no!'

'You yourself have admitted that this case is not a usual one, right.'

'*Sim*,' but did not agree that so drastic a remedy was required. '*Meu deus!* I cannot produce a love interest,' he protested feebly, raising his bulky figure into an upright position. 'Are you mad, Miss Deane? You cannot upset age-old customs.'

'Just what the *Marquês* said, but I would never forgive myself if I didn't make a push to help Leonora. If you can't help me, I'll do something about it myself.' Her mulish look was back.

The doctor gazed helplessly at her. How did one deal with a small, determined *menina*, with accusing eyes that made him feel that he was neglecting his Hippocratic oath? '*Com a breca!* Wait, Miss Deane,' he groaned, scrambling out of his chair with amazing agility in one so huge. 'We must not be too hasty,' patting her shoulder in a fatherly manner.

'Will you do something about it, then? On a doctor's recommendation?' Her wide, friendly smile gathered him into the conspiracy and he was lost.

They had reached the side entrance, when the doctor hastily drew her back, but not before Susan had seen the *Marquês* lift a child from his car.

'The *fidalgo* does not like his deeds of kindness to be made known.'

'Who is the boy?'

'He belongs to one of the families who work on his estate.'

Susan was thoughtful as she drove back to the *quinta*. The *Marquês* could have had an overseer bring in the child, but no, he elected to do it himself. She appreciated that.

★ ★ ★

Wednesday morning came.

The girls and Susan met the *Marquês* as he stepped out of the car. His eyes were on Susan, as fresh and as wholesome as a daisy, in a crisp cotton frock of soft blue that made her hair indescribably fair, a simple straw hat crushing her curls.

'Would you prefer just to take your sisters this morning, *senhor?*' she asked.

'But it was your wish to see the orange blossoms. Remember?' as he saw them all comfortably settled, Leonora in the front while Falina scrambled into the back with Susan.

It was all done with an old-world courtesy that was wholly charming, thought Susan and wondered if his firm mouth had ever softened for a woman. *Senhorita* Pepita perhaps?

The country through which they were passing was flat, with a hint of purple mountains in the distance and would have been uninteresting, but for the immense amount of cultivation. Citrus trees had replaced mango groves and the car slowed down to turn on to a dirt track. Susan sat entranced. The trees were veiled in a mist of white, the green leaves only peeping through here and there.

'We will walk,' suggested the *Marquês*, the heady perfume of blossoms a tangible web and the hum of bees quite audible.

The two girls went on ahead and soon made friends with a black and white terrier. Workers were busy beside the water furrows, the grass thick and lush, in complete contrast to the dry countryside.

The *Marquês* picked a spray handing Susan the delicate offering. 'Girls here wear these at their weddings and there is a belief that couples married at this time are doubly blessed,' which made her wonder if the *Marquês* himself would marry during the next couple of months.

The path had roughened and his hand came up under her elbow, his shoulder close, the half mocking face turned towards her.

'Have you marriage in mind?' The words were innocent enough, but for the wicked gleam that accompanied them, which made her turn hastily away.

She shook her head.

'And what about Mark Fenn?'

'*Senhor*, he's only a friend!'

Susan had gone out a few times with Mark, but only in the afternoons, as she knew he would never consent to bring her home at a reasonable hour.

'Friendship can be a mask for something stronger, especially in a strange land. Two English people amongst a bunch of foreigners — ' his shoulders shrugging with a touch of humour. 'Don't fall in love with him, Miss Deane. He's not the man for you.'

Her hands clenched close to her sides, but she replied quietly. 'He's good company.'

'To get away from the *quinta*?' gentle satire tipping the words.

The conversation was beginning to be too personal and she wished fervently that Leonora and Falina would return, but they were still far up the path.

'Well, Miss Deane, you're not usually so silent.'

How could she tell this man what real freedom it was to be able to step into Mark's car and leave the *quinta* and its problems for a few hours?

'Is it the narrowness of our life from which you wish to escape, or the fact that we don't always see eye to eye?'

'If you must know, *senhor*, yes, I do find it a refreshing change,' deliberately keeping her tone light.

Susan had never been so conscious of a man before and could not analyse this feeling that had taken possession of her. Was it just self-consciousness, this awareness of being completely out of her depth? It was certainly an experience to argue with a man, like the *Marquês*, who was not content until he had unpicked the subject down to the minutest thread. All other men of her acquaintance, would have been highly embarrassed if the conversation had been deeper. Susan was very grateful when Falina came running back to say they had reached the main canal and it was time to retrace their steps. As they walked back to the car, she thought how exactly the *Marquês'* eyes matched the colour of the bulrush heads that grew in such profusion alongside the furrows.

'Now we will pay a visit to my old *ama*,' said the *Marquês*, giving Susan a quizzical glance as she pinned the spray of orange blossoms to her dress.

They drove along another rutted track, until a small, whitewashed cottage was reached. A young woman came out, her dark hair drawn back sleekly, a smile of welcome on her thin face and was introduced as Sondella.

'Please come in. Mother will be pleased to see you.'

They followed her into a tiny livingroom, spotlessly clean, freshly hung curtains at the window. An old woman, in a long black dress and snowy apron, sat in one corner, shelling peas, but placed the bowl aside, with shaky hands, when the visitors entered the room and tried to rise, but the *Marquês* quickly placed a hand on her shoulder.

'You do not have to stand for us,' he said, raising his voice a little and Susan was surprised at the note of affection.

'*Bom dia*,' the old woman said, smiling at Leonora and Falina and the faded brown eyes in the wrinkled face was curious as they turned to Susan, who had the feeling that they could see deeper than most.

'This is Miss Deane, *ama*, who looks after Falina and worries about Leonora.' He turned to Susan. 'Carlotta looked after me when I was a small boy and how strict she was!'

'Don't believe the *fidalgo*,' scolded Carlotta. 'He could wind himself around my soft heart and I was as good as nothing.' Then shot another keen glance at Susan. 'You have a happy face, *senhorita*. Falina and Leonora are lucky to have such a one as you for their *ama*.'

Sondella had brought in the wine and Susan eyed it apprehensively, but tasted it when the *Marquês*, with some amusement, assured her that it was quite mild. It was delicious, cool and not too sweet, the perfect drink for such a hot morning.

Leonora had been strangely quiet, since their arrival, Susan conscious of an underlying strain, but was startled by the intensity of her words, forced through trembling lips, when she said abruptly, 'Please tell me my fortune, *ama*,' and Susan caught the perturbed look the *Marquês* gave his sister. 'I must know now what is going to happen to me,' she added tautly.

# 5

After a swift glance at the *Marquês*, Carlotta took Leonora's hand and peered intently at its lines.

'It's too close to see, *pequena*.'

With a quick indrawn breath, Leonora snatched her hand away and Susan caught the tragic eyes as she slumped into her chair. Surely Leonora was not such a goose to think that Carlotta was afraid to tell her what she saw and was grateful when Falina asked for her fortune to be told.

'Hé, you shall have a gay *cavalheiro*, one that will come on a fine black horse and carry you away from under that haughty nose of your brother.'

Falina's smug expression greatly eased the tension.

'You're forgetting our guest,' mocked the *Marquês*.

Susan drew back quickly. She had no intention of allowing Carlotta to probe into her future, but Carlotta, with a sudden fey look in those faded brown eyes said, 'You do not see the way clearly yet, but the man with the mark will make you forget your independence and only the joy will remain.'

'Will he have to be an Englishman?' queried the *Marquês*, noting Susan's scarlet face.

'You should know the answer to that, *fidalgo*,' gently taunted Carlotta, still with that strange expression and Susan wondered why he should turn away.

'Never that, *ama*,' he replied savagely and without his customary good manners, strode from the room.

Only Carlotta seemed unperturbed by his behaviour, a little smile, almost tender, certainly amused, playing about her thin, wrinkled lips.

Falina could talk of nothing else on the way home, but Leonora sat still and silent next to her brother.

'I don't believe a word of what Carlotta said. Surely you don't, Leonora?' asked Susan, when the silence became too oppressive.

'Why — why couldn't *ama* have told me something, anything. No, all she could say was that it was too close and I don't believe that.'

'Would you accept a fairy tale, like the one she spun for Falina?'

The visit had affected the *Marquês* too, for he remained aloof, intent on the road, but now said angrily, 'Carlotta knows nothing.

Think no more about it.'

Susan was relieved when they reached home. It had been a most uncomfortable visit. As she slipped into bed that night the words of the *ama* came to her again. All fortunetellers told you something similiar, but what really had amazed her, was Raynor Monteiro's reaction and, she remembered drowsily, that he had not asked for his fortune to be told. The *fidalgo* went his own way. Would Pepita share that way?

Susan had just arrived back from dropping Falina off at school when the *senhora* met her in the hall to tell her that they were expecting Nikki, a cousin of the *Marquês'* to pay them a visit.

'Now there's a *maroto*. How you say, a one for the ladies?'

It was evident that the *senhora* was very fond of Nikki, for she became quite animated as she told Susan that he had been to Portugal to visit his parents. 'You will like him, Miss Deane. Hé, what a boy! Our other news is that we are expecting *Senhora* Oliveira and her daughter very soon now. You have heard about Pepita, yes?'

'Mark Fenn told me about her and said that she was very beautiful.'

'Did not the *Marquês* tell you about her?' she demanded. 'You have been in his

106

company several times, surely he must have told you about the woman he is to marry?' and was disconcerted when Susan shook her head.

'I'm sure the *Marquês* would never discuss such intimate matters with me.'

*Tia* Maria shrugged. 'No matter. The *fidalgo's* marriage is an event we have all been waiting for. He's almost thirty and it is time he took a wife.' Her manner suggested that to be a bachelor at that age was a crime against her sex and should be rectified with all speed. 'The final arrangements are to be made when *Senhora* Oliveira returns home,' her sallow face mirrowing intense satisfaction.

Nikki Monteiro arrived that afternoon. He was darker complexioned than the *Marquês*, with well-cut features, an athletic build and looked extremely handsome with a small moustache and liquid Portuguese eyes.

As Susan helped the *senhora* to serve coffee, she was conscious of the flattering glances he threw in her direction and made her escape as soon as she could, leaving the family chattering excitedly about mutual friends and relatives in Portugal.

So that was Nikki and had no hesitation in forming the opinion that he wasn't only a rascal, but a rogue and had every intention of keeping out of his way.

Falina joined her a little later and asked if they could play hide and seek in the garden.

Susan was tip-toeing around a bush, her head over her shoulder, when she found herself being expertly held. It was Nikki, his eyes openly adoring her.

'Where have you been all my life?' he demanded gaily, his teeth very white against his swarthy skin, his polished manner leaving Susan in no doubt as to what he thought of his own attractions, as she hastily extricated herself.

'Not around for your special benefit!'

'Enchanting!' he murmured delightedly, not at all put out. 'I have caught myself a fair princess and I'm your very willing slave, *senhorita*,' his hand going to his heart as he bent low.

'I dislike flowery speeches and flowery gestures,' and would have turned away, but he barred her going.

'*Senhorita*, you are too cruel. Come, here is a seat. You must know we, as a race, very much admire a fair *menina*. We will be friends?' pleadingly.

'No,' she said firmly. 'It wouldn't do at all, *senhor*.'

'But why, Princess?'

'Because you are related to my employer.'

'Only a distant relative,' he amended

108

adroitly. '*com a breca*! here comes my cousin now.'

A clipped voice, with a hint of steel, spoke rapidly in portuguese and for once, Susan welcomed the *Marquês*.

'*Tia* Maria would like to see you, Nikki,' and dismissed the crestfallen young man with a flick of the fingers.

'I'll be seeing you again, *senhorita*.' Nikki smiled, ignoring the frown on the *Marquês*' face, then strolled nonchalantly away.

Susan smiled at the easy way he had taken his dismissal.

'You find him amusing?'

'He's very charming — '

'That would be about right,' conceded the *Marquês*, tight lipped. 'He knows how to pay a compliment, that one. They flow from his lips like drops of honey. Miss Deane, do not encourage him.'

'No, *senhor*,' she replied demurely.

Her reply had, for once, disconcerted him as no hot rejoinder would have done and went to find Falina.

The roses were beautiful and Susan was out early cutting them before the hot sun could wilt them. She was so absorbed in her task that she did not notice Nikki until he was right beside her and looked up startled, to see a thoroughly disgruntled young man.

'Nikki, why the gloom? It's such a lovely day.'

'Life is very bad, indeed,' he mourned, taking the trug from her. 'I've been banished as overseer to a group of allotments five miles from here. I thought Raynor might have been kind and given me a job where I could live here at the *quinta*, but no,' shrugging listlessly.

Susan was surprised at his bitterness. 'Five miles isn't the end of the earth,' she rallied him.

'It's a million miles away when it takes me from you, *senhorita*.'

'Really, Nikki!' she replied with exasperation.

'And what is more,' he went on morosely, 'Raynor has forbidden me to come into Marula unless it is absolutely necessary and that means the *quinta* too. Come out with me tonight, Susan, I go tomorrow.'

His manner had changed to one of passionate entreaty and she wished fervently that she hadn't smiled quite so kindly at him. 'No, Nikki, the *Marquês* wouldn't approve and now I must get these roses into water.'

She met Leonora in the hall and demanded to know why she wasn't entertaining Nikki. 'Here we have a perfectly charming young man in our midst and you ignore him. I don't

110

understand you, Leonora.'

She shrugged indifferently. 'We're not *sympatico* and besides, I've known him all my life. He was a horrid boy.'

★   ★   ★

Susan had just put out her light, when suddenly she tensed. Someone had begun to sing and before she could get up and see who it was, Leonora came in, her eyes dark and wide in the moonlight that flooded in through the french doors.

The man poured out his song, liquid notes of pure magic that floated up into the room.

'Who do you think it is?'

'Perhaps it's the man we saw at the *pousada*, the one who kept looking our way?' whispered Leonora. 'I know he *admired* me, one can always tell,' she added artlessly. 'You — you don't think he's serenading you? Mark perhaps?' a jealous note in her words.

Susan's chuckle was pure mirth. 'Decidedly not Mark,' and felt Leonora relax.

The singing continued, heartbreakingly beautiful. Both girls were now behind the curtain peering down.

'It's not Nikki because Raynor took him to the allotments this morning. *Oh Susan*, he must be singing especially for me. Isn't it ro

111

— *romantic?*' and the two girls stood entranced until the song ended and watched, with regret, the figure melt into the shadows and then he was gone.

'What on earth is your brother going to say?' Susan sat down with legs that shook.

'He's away. Remember?' replied Leonora serenely, taking the question literally.

As Susan drew up the bed covers again, another gurgle of laughter broke from her. This was Doctor Martenis' way of injecting a *little* romance into his patient's life and was envious that she hadn't thought of it first.

She decided that the evening had definitely been worthwhile, for a different girl sat down to breakfast, one that had blossomed overnight.

Leonora ran along the passages as if she had wings on her heels, alive, vibrant, bringing down a sharp rebuke from the *senhora*. Her skin glowed, her hair, that she usually wore on her shoulders, had been piled high on her head, banded by an emerald green scarf, giving her a more mature air.

Susan watched her with mixed feelings. Was this girl to be hurt again, but found to her surprise, that Leonora had decided views on the subject.

'Stop worrying! Of course I know he's only a *cavalheiro*. His song tells me he dare not approach my brother, but he wants me to know that he admires me very much,' her eyes mischievous. 'I find it charming, but I realise nothing can come of it.'

'You will not go out and meet him if he comes again?'

'Of course not!' Leonora was shocked. 'The man I marry will come from one of the best families, either here in Mozambique or Portugal.' This was said without a trace of snobbery; to her it was natural, for she, herself, belonged to an old and noble house.

'Susan,' the wistful child was peeping through again, 'Do you think Raynor will — will make arrangements now I'm so much better?'

'Don't you mind an arranged marriage?'

She opened her eyes wide, a gleam of mischief again between the long, thick lashes, though her words were prim. 'I'm sure my brother will choose a most suitable husband.'

These girls, so steeped in custom and tradition, did they all conform, wondered Susan curiously, or were there some who broke the restricting ties of family to find romance for themselves?

'Perhaps your brother will blame me for not getting rid of the singer?'

Leonora laughed gaily. 'How would you have done that? Thrown a jug of water over him? I would not have allowed that.'

The unknown singer returned on three successive nights, Susan becoming more and more apprehensive as the expected arrival of the *Marquês* drew near. In the darkness, the beautiful, compelling voice vibrated in the still night air, the magic finding a responding echo in her heart. She had never felt so unutterably alone.

The singing stopped abruptly. Susan threw back her covers in time to see the spear of headlights on her bedroom wall and heard the powerful purr of the car stop. Had the *Marquês* seen the figure on the terrace and decided he could not have done so and quickly went along to Leonora's room to find her weeping hysterically.

'Do you think Raynor will call the police?'

'Of course not. If he had, can't you imagine the uproar? Now dry your eyes, you goose,' which produced a watery giggle.

★   ★   ★

When Susan went down to breakfast, Leonora was already there, the glance she threw at her, a little anxious, but the *Marquês* was at his most urbane and talked only of his trip to Beira.

'The singer must have hidden behind a

114

bush,' said Leonora as they went back upstairs. 'Susan, we need not have worried. F — Falina,' she called out buoyantly, 'I'm going riding. Coming?'

A quiver of amusement touched Susan's lips and went along the passage to the sewing room where she had a half finished dress she was making. Suddenly a rose fell close at her feet and Nikki entered, a wide grin on his face.

He was very dashing this morning in a fine, long-sleeved shirt and black slacks. Susan asked him rather crossly, why he was here.

'I told you I would see you again, beautiful,' he said reproachfully, 'and when Nikki Monteiro wants something, he gets it.'

'Why aren't you at work?' trying to ignore his admiring glances, which were most disconcerting and embarrassing.

Nikki had straddled a chair close to her, his arms folded on its back. 'I'm sorry our last meeting was so short, *senhorita*. Oh, how I envy my cousin who sees you all day long.'

'He doesn't see me all day,' she retorted tartly. 'He's a busy man and that is what you should be, right now.'

'But I'm seeing how beautiful you are,' he replied with a wide grin.

Turning away from those humbugging brown eyes, Susan longed to tell him how she loathed his fulsome praise.

'Come, *meu amor*, let's go into the garden,' taking her her hands in a firm grip. 'I have never yet had to compete with a sewing machine. *Com a breca!*'

She wrenched her hands free, but Nikki continued to coax her to go out with him one evening.

'We'll drink *vinho tinto* at the *pousada* and dance. Men will envy me my fair *menina.*'

Susan's tone was repressive. 'I don't go out in the evenings.'

'My cousin can't be that hard on you. On me, yes!' He leaned forward. 'Put on your loveliest dress and we'll dance in each others arms.' Seeing her distaste of the idea, his eyes rapidly turned from soft coaxing brown to the hardest of road pebbles, his voice harsh. 'But I insist!'

Falina coming in at that moment, seeing his face, screamed. The *Marquês*, whose study was just below, must have heard her for seconds later he came into the room.

'Nikki, what are you doing here?' he asked with ominous calm.

Nikki's eyes dropped under that angry gaze. 'Why am I not allowed more freedom? I had to bring a man in to the hospital,' he muttered mutinously.

'Was there no-one else?'

An embarrassing silence fell as the *Marquês* waited for a reply, but none was forthcoming. Nikki's face had gone sullen and Susan suddenly felt sorry for him. She quickly slipped from the room taking Falina with her, but not before she heard the *Marquês*, say in his coldest way. 'It is time you did an honest day's work and if you don't improve, I'll have no option but to send you back to your parents. Now you may go.'

Seeing Susan later that day, the *Marquês*, with a mocking twitch of his mouth said, 'Don't fall in love with him, *senhorita*.'

'I have no intention of doing that. He's very young, isn't he?' A shiver feathered Susan's spine, for there had been in his manner this morning, an unexpected streak of ruthlessness when his will had been crossed. Not a little boy crossness, but a far stronger reaction.

The *Marquês* chuckled unexpectedly. 'Poor Nikki, my heart is wrung for him! What an insult to his blossoming manhood.'

'Please don't tell him I said so,' she asked anxiously.

★   ★   ★

The *Marquês* and Leonora had left for Beira the previous afternoon and now the whole *quinta* was relaxed. Even the old *senhora* was lingering over her coffee this morning, which they were having on the patio, where the bouganvillaea were especially beautiful and Susan never tired of their long spears of colour, red, orange, mauve and pink, arching to the ground.

The *Marquês* had invited Susan to accompany them, but she had declined, saying that it would do Leonora all the good in the world, to have her brother to herself for a few days in the stimulating atmosphere of one of the very fine hotels in Beira.

Leonora had been radiant, her pink frock with its black accessories, suiting her dark colouring admirably and it was hard to realise this was the same girl who had lain on her couch, listless, bored and unhappy, not so long ago. Even her brother had admired her as she stood framed in the oak doorway and Susan had felt a thrill of achievement at the better relationship that now existed between the two.

The doctor had been invited to lunch, just before the *Marquês* and Leonora had left, giving Susan conspiratorial glances that spoke volumes, his arched, bushy eyebrows and inclinations of his head, making her smile back impishly at him. It was a silent tribute to their joint effort, but she hoped fervently that they had seen the last of the singer.

# 6

Susan had unending admiration for the way *Senhora* Ferreira ran this huge house, marshalling her servants as ably as any general, but she had never become friendly with the old lady, always sensing a faint antagonism directed to her. The *senhora* never openly showed it in front of the *Marquês*, but was not above letting her feelings be shown when she and Susan were alone.

Timbo removed the coffee tray and still the *senhora* lingered. 'You like the *fidalgo*, Miss Deane?'

'I love working here,' she replied noncommittally.

'Do not become too fond of him. I notice how he talks to you and I think he admires the way you have with his sisters,' her beady eyes watchful.

'The *senhor* is my employer.'

'What matter? He has great charm and wealth as you must have guessed,' waving expressive fingers. 'I was not at all happy when Raynor told me, you would take Miss Brack's place.' Her smile was cold. 'You are far too young and young girls are impression-able.'

Susan had difficulty in keeping her temper. 'It is understood that I leave as soon as I have found out about my brother, or Miss Brack returns.'

'Ha, your brother!' The pouchy body straightened up. 'I had almost forgotten your reason for coming to our country. It must be a sadness to you to know that he was mixed up in a drunken brawl.'

'Peter wasn't like that,' replied Susan vehemently. 'Did you not meet him?'

'One can be mistaken in a person,' *Tia* Maria murmured, clasping podgy hands in her ample lap.

'Well, I'm determined to find out what really happened.'

'Perhaps it's not always wise to probe too deeply into such matters, *senhorita*. You could uncover something worse.'

The appearance of Luiz saying that the *senhora* was wanted on the 'phone prevented Susan from replying. She had just reached the hall when *Tia* Maria came out of the study and beckoned her to the instrument.

'The Post Office say there is a cable for you from London.'

Susan picked up the receiver and asked that the cable be read. It was from Nannette saying she would be arriving in Beira on the 26th.

Susan slowly replaced the receiver. She hadn't even thought of the possibility that the message might be from her sister-in-law and, certainly, not to say she would be coming out to Mozambique.

'Not bad news?'

'My sister-in-law arrives on the 26th,' Susan was still a little bewildered.

'You'll be booking a room at the *pousada*?'

Susan gathered her wits together. 'Yes, and when is the 26th?'

'It's tomorrow. What a pity you did not know sooner. Raynor could have met her and brought her back. Would you like me to 'phone him?'

Susan nodded gratefully and the *senhora* promised to 'phone at lunchtime.

Susan had just reached her room for the siesta, after having settled Falina, when the *senhora* entered, kneading her hands in agitation and saying that she had been unable to contact the *Marquês*.

'You'll have to go and meet this relation yourself.'

'I'll leave straight away,' replied Susan opening her wardrobe. 'Please tell Falina, I'll be back sometime tomorrow.'

*Tia* Maria nodded, Susan missing her smug expression.

★ ★ ★

Ten minutes later Susan was on the road and found the prospect of visiting Beira again, exciting, but what had decided Nannette to come out here? That was the question that occupied most of her thoughts as she passed small towns, deep dropping valleys and sudden openings in the bush that revealed clumps of banana trees or railway sidings.

It was dusk when Susan booked into the hotel nearest to the airport and was grateful to the *senhora* for the information she had given her, but why had the old woman veiled her glance so hurriedly when she had suggested that she meet Nannette? There had been more than a gleam of satisfaction in those dark brown eyes, which Susan had begun to distrust, then smiled to herself. The *Marquês* was going to be furious. Perhaps that was what the *senhora* was hoping for?

Her case was taken to a tastefully furnished room on the first floor and as she stood on the balcony, enjoyed the cool night air, listening to the sea crashing on an unseen shore.

She was sitting in the lounge after dinner,

123

reading a book, when someone slipped into the chair next to her.

'Nikki!' she gasped. 'I'm sure you shouldn't be here.'

His eyes danced with mischief. 'There are always those who are willing to do other's work for a few *escudos*.'

'That's asking for trouble!'

'Don't be so uneasy for me, but I find it very sweet that you should care. Now I have you to myself for the whole evening.' He leaned forward, nearly touching her. 'Did you know your eyes were blue pools of mystery that I would very much like to drown in?'

'How did you know I was here?' she demanded angrily. 'No, don't tell me. It was *Tia* Maria.'

Susan was sickened at all the duplicity. 'I suppose you thought it quite safe, knowing the *Marquês* was away? It is time you grew up, Nikki,' her tone severe and was not really surprised at his reaction. She should have held her tongue.

His thick, black brows met over hot brown eyes, a sneer on his thin lips and Susan shrank back.

'Always, always I've had to live in the shadow of the great *fidalgo*,' his clenched fist smiting his palm with violence. 'Why wasn't I born to be master of that great *quinta*? No,

I'm only a twig of that noble family,' the words coming out in a bitter torrent. 'He's a fool, that one. He should enjoy his wealth.'

'You don't have to stay and work for the *Marquês*.'

'No,' he conceded, pleating his bottom lip.

'Well, go on and say it,' she invited derisively. 'No other employer would be so lenient with you. I know!'

'This is a silly conversation,' he said leaning closer, his bold glance lingering on her face. 'Let's go and dance, somewhere. You don't have to consider all our stupid conventions.'

Susan stood up as he burst into impassioned Portuguese and although she only understood the odd word, it brought a wave of hot colour to her cheeks.

'A thousand pardons, *senhorita*, but it is so much easier to talk of love in one's mother tongue. I get drowned in the English.'

'It's just as well. Now, if you'll excuse me,' turning away purposefully, ignoring his pleadings and then remembered she had never asked him about Peter. 'By the way, did you ever meet my brother?'

'I did meet him, yes.' He drew hard on his cigarette, his fingers not quite steady. 'I know Raynor was very upset when he heard that he had died in a fight. I feel for you, Susan,' but his eyes refused to meet hers.

She turned away. 'I was hoping you could throw some light on the fight that night in Marula,' but he was not listening.

'*Com a breca!*' he muttered savagely. 'Excuse me!'

Susan stared at his fast moving back, making for the french doors that led on to the verandah. Whatever had made Nikki leave so abruptly and walk across the room towards the reception area, then saw the reason why Nikki had faded so neatly out of the picture.

The *Marquês* stood there, a stony brilliance to his dark eyes, his anger the litheness of a panther, rippling below the surface, his eyebrows flared with startled disapproval.

How very unchivalrous of Nikki! she thought indignantly.

'Well?' enquired the *Marquês* coldly. 'You refuse my invitation to come to Beira and yet I find you here?' I saw my cousin disappear, but it was only when you turned around that I recognised you.'

Susan felt her mouth go dry and the colour fading from her face, but her eyes never wavered from his anger.

'I would remind you, *senhor*, that I may do as I please. I'm entitled to some time off, surely?'

'And I would remind you, *senhorita*,' he took up quickly, 'that you're living under my

126

roof, therefore my responsibility.'

Susan had never seen him so tight-lipped before, as he added icily, 'It is not right that you should be here by yourself.'

'What isn't? My being down here because I happen to work for you? I'm sure if I was just another tourist, you would not adopt this archaic attitude,' she replied calmly.

The *Marquês* had become aware of the bold and interested stares of the other hotel guests. With a slight bow and still coldly formal, he took her by the arm and led her out into the softly lit garden. 'When you are with us, *senhorita*, I would ask that you conform and respect our customs, as I have said before.'

'Even telling me who I can't fall in love with, or, arranging a marriage for me?' she asked and knew a pang of shame. 'I'm sorry, *senhor*.' The expected flare of anger did not come.

'Even that, Miss Deane, if you should desire it,' his mouth taking on an upward tilt and she knew he was laughing at her.

How overbearing could a person become and how insufferable! To think that she had woven dreams around this man, building up a hero image, fostered by Peter's letters and now reality had fallen far short of that picture.

They were sitting on a swing seat, the *Marquês* in a white dinner jacket, making him appear even taller than he was, the arrogant nose and chin lifting in a manner that her mother would have termed 'about himself', but his eyes had begun to twinkle and Susan's anger receded as quickly as it had been born. Whatever else she might think of this man, there were moments when he could be charming.

'Circumstances can appear deceptive, *senhor.*'

'True,' still mocking her. 'I'm willing to listen, Miss Deane. Perhaps, now that your pride has been appeased, we'll get to the truth of the matter?'

Susan glared at him again, disconcerted to find him so discerning and told him of Nannette's impending arrival.

'And Nikki brought you down?'

She shook her head. 'He only arrived after I'd had my dinner. You shouldn't jump to conclusions,' she added severely.

Was there ever such independence in one small person, thought the *Marquês* with exasperation. Not to his knowledge and it was new to him. The cold anger at finding her here alone, drained away, but disapproval still remained. 'I don't like the thought of you coming all that way by yourself. Anything could have happened.'

There was no pleasing the man. If she had come down with Nikki, he would have been furious.

'I enjoyed the trip.'

'But surely, other arrangements could have been made?'

Susan told him how the *senhora* had rung his hotel but had been unable to contact him and that it had been her suggestion she meet Nannette.

'Do you know which hotel she rang.'

'No, but she did say it was the one where you usually stayed. Are you booked in here, *senhor*?'

'No, I came to see a business acquaintance.'

'But where is Leonora?'

'With friends,' his smile warm. '*Senhora* Attunes would not hear of her staying at the hotel with me. Perhaps it is just as well we now have an excuse to go home tomorrow. I've had qualms about my bank balance several times these last few days. I think Leonora has shopped in every store in town.'

An answering smile crinkled the corners of Susan's eyes. 'But I thought you were only going home at the end of the week? I can easily drive my sister-in-law up to Marula.'

'You and *Senhora* Deane will drive back

with us and why not tomorrow?' he asked smoothly.

'But what about my car?' It had suddenly become a vital part of her, representing one small part of independence left to her.

'I'll have it driven up to the *quinta*,' he replied easily as if it were a very small matter indeed. 'And now I think you should go to bed. The 'plane comes in very early.' His glance lingered on her face and then said brusquely, 'I don't like to think of you alone, *pequena*. You'll lock your door. Is it understood?'

They were walking along the luxuriously carpeted passage, stopping outside Susan's door. She chuckled up at him. How very Portuguese he was and for the first time, it gave her a warm feeling to know that someone cared enough about her well-being and was concerned for her safety.

'Nothing possibly could happen to me here, but I promise to lock my door,' she added hastily seeing the smouldering light back in his eyes. 'Goodnight, *senhor*.'

She held out her hand to him and was startled when he gently kissed her fingers; heard his murmured '*boa noite*' then shut her door and locked it.

Don't be such a fool, Susan Deane, she admonished herself severely, finding that simple gesture had shaken her more than she wanted to admit. That kind of salutation was as old as the hills and perhaps it was his way of apologising.

\* \* \*

Susan was up early the following morning and rushed on to the balcony to get a glimpse of the sea. A wide expanse of white, sandy beach met her delighted gaze, huge waves crashing down in high curls of aquamarine.

The *Marquês* came into the diningroom just as she had finished her coffee, waving aside her apologies.

'You are not late, Miss Deane, I'm early. You slept well?' he asked politely, his eyes resting on her youthful face and smiled. 'I see you did.'

How well this English flower had been transplanted, he thought. Her skin had drunk in the sunshine and her hair was bleached even fairer than it had been on her arrival.

'What happened to Nikki?' She couldn't help asking.

'He's gone back.' The closed look prevented her from probing further. 'If you are ready, Miss Deane, we'll go.'

They approached the airport, holiday chalets fringing the seashore, coconut palms everywhere and were in time to see the 'plane land.

Susan, from the balcony, watched Nannette walk across the tarmac in a figure-hugging snuff brown suit with white accessories. She was talking gaily to a man who was carrying her coat and cosmetic case. He had eyes only for her.

Custom formalities over, Nannette came out to greet her sister-in-law with hands outstretched.

'Susan! Wonderful to see you again,' but her glance was only for the *Marquês* as she waited, expectantly, to be introduced. He bent over her hand and then acknowledged her introduction to her companion, who still hovered restlessly beside her.

'You'll come to Beira soon, *Senhora* Deane?' pleaded Alberto Lopez, his short-sighted eyes peering at her, through thick spectacles as if trying to imprint her image on his memory. 'My mother will be pleased to welcome you.'

Nannette gave a soothing reply and the little man disappeared into the crowd.

'I'm so glad I decided to come to Mozambique,' turning to the *Marquês*. 'Portuguese men are simply wonderful. They make a woman feel so desirable.'

The *Marquês* inclined his head. 'A pity you

did not think to come out sooner, *senhora*,' he replied pointedly, opening the car door and handing the two girls into the back seat.

'I just couldn't possibly have come out with Susan. There were engagements that I couldn't break.'

The *Marquês* dropped them at the hotel and said he would pick them up at ten o'clock.

'What was that all about?' asked Nannette with lively curiosity, throwing her hat and bag on to the bed.

'I drove myself down yesterday, but he has made arrangements for us to go back to the *quinta* with him and his sister.'

Nannette stared at her, a calculating gleam in her dark blue eyes. 'Darling, couldn't you work it so that you took the sister, leaving me to go with that very presentable man?'

'You don't know the *Marquês*,' said Susan with feeling. 'What he says goes. Did you have a good trip?'

'It was all right, but Sue, why didn't you mention that Raynor Monteiro had a title?' she asked reproachfully.

A warning pulse throbbed in her throat. 'I didn't think you'd be interested,' hedging. 'What about that man of yours in the Stock Exchange?'

'Not interested in a *Marquês*?' Nannette

was horrified and strolled on to the balcony. 'Gorgeous,' she said indifferently, turning back into the room. 'I must take a shower. Have this pressed for me,' taking out an amber sheath from her case. Susan rang the service bell.

'How on earth do you cope with makeup in this heat? I look positively ghastly.'

'I don't.'

'So I noticed.'

'At least I know it hasn't streaked,' untroubled by her sister-in-law's remark, but Susan could see nothing wrong with that beautiful face pouting in the mirror. Wide set eyes as blue as a summer's day, dark brown hair with a hint of russet.

'Nannette, why did you come out here?'

'To meet the owner of the *quinta*, of course. It might be quite interesting to have a title,' she said airily.

'But he might have been married, for all you knew.'

'Darling, you would not be working for him, if that were the case,' her tone shrewd.

The hiss of water made further speech impossible. Susan did not like to discourage Nannette in her thinking, she might even wonder if she was interested in the *Marquês* herself, but the mere fact that he was still single, proved that he was not a man to be

134

easily captivated by a pretty face. He would do his own courting.

'I think you should know that the *Marquês* is interested in a Portuguese girl, a very beautiful one too, by all accounts. She and her mother have stayed at the *quinta* and it is only a matter of time, so I believe, before the engagement is announced.'

The maid brought back Nannette's dress.

'Men have been known to change their minds,' she replied, slipping the frock carefully over her head, her eyes startling blue above the amber silk.

'Oh, Nannette, why didn't you come out here while Peter was alive?' Susan asked sadly.

'Have you been able to find out anything about Peter?'.

Susan shook her head and went out on to the balcony. The street below was now thronged with tourists in gay beach wear.

'Good. That's the reason why I have come to Mozambique. To find out what really happened to my husband.'

\* \* \*

The *Marquês* drew up punctually, in front of the hotel at ten o'clock. Susan had become more and more exasperated as Nannette dawdled over her dressing.

'Don't panic, Sue. Keep them waiting, that's my motto. Men appreciate you more that way.'

'I call it plain rudeness,' she retorted and left her sister-in-law to complete her packing.

Leonora was overjoyed to see Susan again and talked of all the fabulous dresses she had bought. Susan glanced laughingly at the *Marquês*, who raised eyes and hands in mock protestation.

'If I'd known what a difference clothes could make to a woman, I'd have brought her down long ago.'

Nannette chose that moment to make her appearance, appealingly fragile, a suitcase in each hand. She could have called a bellboy, thought Susan with wry humour.

The *Marquês* hurriedly relieved her of her cases, introducing her to his sister.

They drove away from the coast, Nannette in front, demanding a running commentary on the scenery and on the groups of laughing African women, carrying huge bundles of wood on their heads, their babies strapped securely on their backs.

'Quaint dress,' she remarked.

'Two tablecloths,' murmured Leonora.

This was, indeed, what the women were wearing. One twirled around their waists, the other tied under their arms. They made a colourful scene as every cloth seemed more gaudy than its fellow.

'I'm pleased you have eventually arrived in our country, Mrs. Deane,' said the *Marquês* politely. 'We have something here in Mozambique both exhilarating and challenging,' waving a hand at the vast stretch of cultivated land through which they were passing. 'A rich heritage to pass on to one's son, do you not agree? We are not as backward as you might have supposed.'

'Oh, it's all so different from anything I've ever seen,' enthused Nannette, leaning a little closer to his shoulder.

'Any nation that keeps such a strict watch on their womenfolk is backward,' retorted Susan, who, for some unknown reason, was decidedly ruffled and even Leonora gave her a startled glance.

The *Marquês* smiled, his eyes meeting hers in the rearview mirror. Susan glared back at him, but he refused to be drawn. 'We are a hotheaded race, *senhorita*, so there must be some form of discipline, so we guard our daughters and after all, there's much to be said for arranged marriages, as I've said before. At least the couples are spared the pain of jealousy and uncertainty.'

He spoke with a moroseness that made

Susan look at him quickly, but was unable to see his face.

Nannette's tinkling laugh broke the tension. 'By the way, Susie, Felix sends his love.'

'The *senhorita* is not a Susie. It does not suit her.' The words had come out sharply which made Nannette's cheeks hot.

'I didn't think he would have even remembered my existence,' said Susan hastily. 'Did you know that Mark Fenn was out here?'

She stiffened. 'Such a lot you missed out in your letters, darling. He's charming, of course, but I don't think he's the man for you.'

'I like him. He's a friend.'

'But much too old in experience for you.'

Susan was just about to refute this, when they pulled up at a garage. What did it matter anyway, for she was sure Raynor Monteiro was not a bit interested in her past, but she wanted, perhaps a little irrationally in view of this disinterest, to protest that Felix was no boyfriend of hers.

They were on the road again, the car picking up speed rapidly, the *Marquês* once more relaxed behind the wheel. He drove as if he were part of this powerful machine.

How arrogant and feudal he could be, but how very thoughtful, on occasions, and her sense of grievance melted away, leaving something humbling and strangely exalting. Now she knew why she had been so prickly and on the defensive all morning. It had needed the arrival of her sister-in-law, to make her know her own heart.

Nannette was sitting even closer to the *Marquês*, her fragrant perfume wafting to the back of the car. Susan was aware of the beat in her throat, knowing a surge of joy and pain. The *Marquês* was not for her, for his destiny and hers would never meet. He owed that to his name. It was better to stifle this love now at birth, for she knew if she allowed it to grow, it would devour her.

They had reached the outskirts of Marula, Susan informing the *fidalgo* that she had booked Nannette in at the *pousada*, but instead of stopping, he invited Nannette to lunch at the *quinta*.

Falina threw herself into Susan's arms the minute the car stopped, sobbing hysterically that she thought Susan had gone for good.

'But darling, didn't you think to look in my wardrobe? No woman leaves without her clothes. Didn't *Tia* give you my message?'

The little girl shook her head, the brown eyes drowned with tears. 'She just said you'd left for Beira, nothing else,' the sobbing easing a little.

'Let's take Mrs. Deane up to my bedroom, shall we?' suggested Susan quietly. 'I'm sure she would like a wash after her journey,' and chalked up one more black mark against the senhora.

'What's this pousada you talk of?' demanded Nannette, as they made their way upstairs.

'It's a local Inn. It will be a bit noisy in the evenings, I'm afraid, but it is the best Marula can offer.'

'But surely, in this huge house, you could have seen to it that I stayed here?' her voice was shrill.

'The senhora did not offer to have you and after all, Nannette, I'm only an employee here. Come and have a wash,' Susan suggested soothingly.

While Nannette was renewing her makeup, Leonora pulled Susan into her room to show off all the clothes she had bought and also to tell her that she had met that nice young man they had seen at the pousada, the night Raynor had taken them all to have coffee.

'I was not able to talk to him for long, because Senhora Attunes wanted to introduce me to her other guests. He's so handsome, Susan and I liked him the best.'

At luncheon, Susan, happy to be back at the quinta and that, at last, Leonora was a normal, happy eighteen year old, did not notice the piercing glances thrown at her by the *senhora* as the meal progressed. *Tia Maria's* shrewd, button eyes, were not slow in seeing beyond the obvious as they stared, wide with chagrin, and dismay at Susan's glowing face. She had been most upset and surprised at seeing them all get out of Raynor's car. It had not been as she had planned at all and where was dear Nikki?

At breakfast the following morning, a message was delivered to Susan, saying that the *Marquês* would like his sisters to accompany him that afternoon, to visit an old friend, who lived on the other side of Marula.

'This invitation includes you, *senhorita*,' added Luiz, turning to Susan with a pleased bow.

She had become very fond of the old butler and he, in turn, paid her a deference that was not merely good Portuguese manners.

'The *fidalgo* also said that if you'd like to invite the *Senhora* Deane, it would be in order,' marked disapproval in every line of his sagging body. He had not liked that *senhora*. Those glances had been too calculating in the way they had rested on the *Marquês*. Luiz had not thought it at all fitting. Old he might be, but nothing escaped his still sharp eyes.

Susan 'phoned Nannette to find that she

had made other arrangements.

'Darling, what a blow! I'm meeting Mark. Wonder if I could put him off?'

So the news of her sister-in-law's arrival had already percolated to the dam site, or had Nannette rung up Mark, wondered Susan.

'Don't do that,' she said quickly. 'It would be a terrible disappointment for him. Every time we've gone out, all he could do was talk about you.'

'Perhaps you're right. Please offer my apologies, but Sue, don't tell him I'm going out with Mark.'

# 7

Susan changed into a frock of lime green cotton and slipped into low heeled sandals. She brushed her hair until it shone and used her favourite coral lipstick, then hurriedly went to help Falina with her buttons.

Leonora had chosen one of her new dresses, a thick white linen with a bright red scarf cunningly knotted around the neckline, but the *Marquês* did not even notice these feminine niceties, which was unusual, for he always noticed small details, his dark eyes appreciative, though he might not always make a comment. Today however, he was preoccupied and silent as they drove down the avenue and then on towards Marula.

Susan had seen Mark earlier that day and had smiled widely at him as he strode down the *Avenida* Monteiro. He looked happier than she had ever seen him.

'Come and have a coffee,' he invited. 'I'm meeting Nannette at twelve,' and was not put out when she excused herself.

He wore a white shirt in place of the usual khaki one in honour of the occasion and even his hair had been plastered down.

The *Marquês* drew up in front of a small, whitewashed house with a wide verandah, rich with flowering shrubs, which Susan noticed particularly and after the introductions said to *Senhor* Cruz, that his collection of variegated leaf plants were lovely.

'Ha! You like flowers, *senhorita*,' a pleased smile creasing his wrinkled face. 'But come, I must not keep my guests in this hot sun. We'll sit in the garden.' His white eyebrows were like a boss over the dark smiling eyes.

It was very evident that their host was delighted to welcome them and as they rounded a corner of the house, a young man rose from a seat and came towards them. It was the man they had seen that night at the *pousada* and whom Leonora had met in Beira. Susan could feel the start in her as *Senhor* Cruz introduced him as Enrico Ferens.

Cool drinks, in frosted glasses were brought out by a smiling housekeeper and *Senhor* Cruz explained the presence of his guest.

'I used to be a colleague of his father's. A long time ago,' he smiled, 'and now Enrico comes to pay me a visit.'

Susan shot a questioning glance at the *Marquês*, but his face was inscrutable as he passed her the plate of small cakes. His poker face, she thought disgustedly and longed to know whether this was an arranged meeting. It did seem too much of a coincidence that this same man had been the only other person on the balcony of the *pousada* that night.

Susan watched Leonora demure and faintly smiling, sip her drink and afterwards, Enrico's slight inclination of his head in the *Marquês'* direction. He must have asked permission to take Leonora around the garden, for a few seconds later, he went over to her and led her off to see *Senhor* Cruz' very fine collection of birds.

Falina wanted to go too, but the old man smilingly ruffled her dark, straight hair and said, 'I have a surprise for you, *pequena*. Come with me.'

They were back so soon that Susan had no chance to ask the question she longed to know, Falina cradling a diminutive black and white kitten.

They talked of many things that afternoon, the *Marquês*, *Senhor* Cruz and Susan. The young couple were never quite out of sight and Falina was quite happy playing with her kitten, which she had been told she could take home with her.

Enrico bent low over Leonora's little hand as he handed her into the car and when he came over to the driver's side, the *Marquês* asked him to dine with the family the following Tuesday.

Leonora was strangely silent on the way home, not the sullen silence so often associated with her, but a deep contented stillness.

Her brother glanced keenly at her once or twice during the drive, but said nothing. It was only when he turned into the *quinta* avenue that she said in a tone of wonder, 'He — he liked my stammer and said it was charming. Was it not strange that he was the man who was at the *pousada*, Susan?'

Strangely enough, Leonora did not seem to want to know whether this man was to be her *noivo*. Evidently that would have been thought forward. If it had been me, thought Susan, I'd have demanded to know immediately. Even during the days that followed, Leonora evinced no curiosity, content with her dreams.

She went around in a daze, late for meals and Susan was sure, had not looked beyond the immediate future or the possibility that Enrico might not be for her. It was left to Susan to do the worrying. Leonora was obviously in love and this was the danger of this type of upbringing. Coop up a young, impressionable girl and then suddenly introduce her to a personable young man, the obvious must happen.

The *Marquês* was no help either and remained his most urbane, imperturbable self and when he did catch her exasperated glances, returned them with a smile of polished charm.

How dare he move us all like chess pieces, she fumed. He had talked so glibly about the lack of uncertainity in arranged marriages. If all this wasn't uncertainity, she wondered what was.

With the advent of Enrico Ferens, the vexed question of the singer once more rose to the surface and she mentioned the fact to Leonora who was quite unconcerned.

'I — I couldn't help it if some *cavalheiro* came singing at my window,' she said, with a saucy lift of her dark brows, 'but if it will make you feel better, Susan, then speak of it to Raynor. I — I don't mind.'

Susan got to the study door several times that day, but on each occasion, turned away and went upstairs instead. Perhaps she would wait a while and hope no garbled account of the incident reach the *Marquês*. This was cowardly, but the thought of his anger filled her with dismay.

* * *

The weeks flew by and then one morning Leonora came flying upstairs to announce happily, that she was betrothed. 'And so now I have a *noivo*,' a glow softening the dark eyes. 'Thank — thank you, Susan. You have made this possible.'

'You're not just in love with love, are you?' she asked anxiously.

'Of course not. What makes you think that?'

'What about the singer? Weren't you a tiny bit in love with him?'

'I liked the idea of a serenader under my window, yes. It was romantic and perhaps I was in love with love — a little, but can't you see, Susan this is different. I want to do everything for Enrico.' The words were strangely humble.

*Tia* Maria panted into the room, tears running down her face, as she clasped Leonora to her ample front.

'Ha! A celebration we shall have, *pequena*, that will throw all celebrations into the shade,' she promised, already thinking of the families that would be invited, the food she would serve. The servants would start cleaning the huge ballroom again and bustled away happily.

One afternoon Susan plucked up enough courage to knock on the study door.

'Are you busy, *senhor?*'

'If you wish to speak to me, no, Miss Deane,' and pulled out a chair for her. Her consternation must have been evident for he said sharply, 'Something isn't right. What's the matter?'

'There was a singer who came here while you were away.'

The *Marquês* sat back in his chair, a smile quivering the corners of his mouth.

'You don't mind, *senhor?*' a relieved sigh escaping her.

'He was serenading you?' and seeing the shake of her head, went on, 'I like your honesty, Miss Deane.'

'Well, I did feel bad about allowing it, but you see, *senhor*, I had never come across a situation quite like it before and did not know what to do.'

'There's always a jug of water to damp a young man's ardour,' he reminded her with a smile.

'You're not angry about this, *senhor?*' she asked with surprise and relief.

'My dear Miss Deane, I knew about it.'

Susan forgetting her position, flung her arms around his neck, thanking him fervently; the startled *Marquês*, now on his feet, instinctively grasped her to him, before she could spring back, blushing to the roots of her hair.

'*Senhor*, please forgive me,' hastily pulling away.

'Please don't apologise. I found it most delightful. There seems to be much in your way of life that I have missed.'

But Susan wasn't mollified. '*Senhor*, I think you could have told me. I have been very worried.'

'You take other people's trouble too seriously, *pequena*. I like that. Forgive me?'

Susan's indignation subsided. The *Marquês* in a conciliatory mood was hard to withstand. 'Did the doctor tell you?'

'Yes, he came to me in a great state of agitation, saying that if we didn't do something — I think he called it 'a love interest for Leonora' — you would. The good doctor also said, with much perturbation, that he had never before had to deal with so strong minded a *menina*. So, you'll forgive him for telling me?'

'I suppose he was afraid of you?'

His brows came down in two haughty lines. 'Pardon?'

She was covered with confusion. Her wretched tongue had run away with her again.

'No, he was not afraid, Miss Deane, but he has a fondness for me that would not allow

150

him to do anything behind my back. It was a delicate situation, one that you did not fully understand. I think we handled it very well,' his good humour restored.

'Did you have to pay that young man money to keep his mouth shut?'

'Not at all,' he gently mocked. I can tell you now that it was Enrico himself.'

Susan blinked at this second surprising statement. 'So you had made arrangements for Leonora,' she said softly.

'Of course,' leaning forward, 'but I must thank you, *senhorita*, for what you have done for my sister.' Gone was the mockery and in its place, a warm friendliness. 'I had hoped, when you arrived so unexpectedly, that you might be good for Leonora, being so much nearer her age. Miss Brack is an excellent woman, but a little old and see, I've been proved right.'

'But Enrico hadn't seen her by then.'

'He saw her in town one day and we'll keep this little secret from Leonora. She will make Enrico a good wife, but I think it will be of benefit to let her continue to think that there was another interested enough to sing love songs under her window, especially for her. We'll let her keep her little romance. Enrico will not tell her either. We have decided.'

Susan's thoughts were in a turmoil. How

very intimate a knowledge he had of women. Leonora would look back, with a thrill of pleasure as she remembered the man who had serenaded her in a scent filled garden and in some unfathomable way, it would deepen her love for her husband.

Of course, Nikki had been told of the bethrothal, which gave him an excellent excuse to visit the *quinta* to offer his felicitations to his cousin and see Susan.

'Olá, *senorita*,' he said gaily, bringing out his old plea.

'Really, Nikki,' she said with exasperation. 'Can't you take 'no' for an answer?'

'A lot of women say 'no' when they mean 'yes',' he grinned impishly. 'They all like a little gentle persuading.'

'I'm not in that category,' she replied coldly and got up from her chair.

'Just tell me when I can come and fetch you,' he cajoled, laying a detaining hand on her arm.

She remained silent and Nikki's attitude hardened. 'I know! It's 'the *Marquês* doesn't like this, the *Marquês* doesn't like that',' he mimicked. '*Meu deus!* Why should he have everything?' he muttered wrathfully, but he allowed her to pass.

How volatile these people were, thought Susan raggedly and met the cold eyes of the

*Marquês* as she entered the hall. No doubt he was displeased at what he thought was her encouragement of Nikki.

The following morning the *quinta* exploded into excited life. A telephone call had been received, just after breakfast, announcing the arrival of *Senhora* and Pepita Oliveira.

The guest suite was made ready and just after half past four, a chauffeur driven car swept up the avenue and the whole family went out to greet their visitors.

Susan, sitting in the garden, could see the little group and saw the *Marquês* smile as he bowed to the two ladies. He took Pepita's hand in his and raised it to his lips.

'Welcome to the *quinta* again,' she heard him say, with a flattering reluctance to release the little hand that had been held out so appealing to him.

Susan could stand no more and flew up the stairs, biting back the tears that threatened to overflow. She was thankful that she had, for Luiz came in to say she was awaited downstairs and was surprised when the *Marquês*, after introducing her to the two ladies, asked if she would join them, which did not suit the old *senhora* at all.

*Senhora* Oliveira was in her mid forties, tall and thin with slightly greying hair, but with

Pepita one had to take a second look.

Dark, silky hair, drawn back into a french knot, from an oval face; creamy complexion with intense brown eyes that glowed through incredibly long lashes and that passionate provocative mouth? What man could resist this beautiful, alive creature and what a fitting *Marquêse* she would make, thought Susan feeling suddenly awkward, as Pepita's velvety glance skimmed her from head to foot, without missing a single detail of her plain linen frock, then turn away as if satisfied.

★ ★ ★

Raynor Monteiro seemed to have shelved all business during the days that followed as he took his guests through the vast Monteiro estate.

Susan heard all about these trips at meal times, for Pepita could talk of nothing else.

Occasionally the visitors joined Susan and Falina on the patio and on one such afternoon, *Tia* Maria took *Senhora* Oliveira to see her very fine orchid collection. It was evidently an opportunity Pepita had been waiting for, for she asked Falina to go too. The little girl trotted off obediently at a nod from Susan.

Pepita settled back more comfortably in

the round basket chair, her body gently taking its curve. 'I do so love this climate. I much prefer it to Europe, for how can one look really beautiful when one is chilly?' Her accent was delightful. 'Perhaps you think otherwise?'

'I find this a lovely country,' replied Susan warmly, busy knitting a jersey for her namesake.

'But I hear you are to leave soon,' the well modulated voice flowed on. 'I think it is just as well, for I hear that poor Nikki is quite infatuated with you.'

She leaned forward and threw out an appealing hand. 'Please take a friendly hint, Miss Deane. It would not do to make plans in that direction. We know his father well and he has other plans for his son.'

Susan's reply was cool. 'Nikki, as far as I'm concerned, is only a hot-headed boy. I've certainly no intention of losing my heart in that direction, if that is what you mean.'

'I know how you have helped Leonora, but nobody is indispensible,' Pepita replied silkily, shrugging her shoulders under a fine paisley blouse.

'I have known all along that this post is temporary. As soon as Miss Brack returns, I'll leave.'

'Perhaps Miss Brack's services will no

longer be required either,' she said archly.

'In that case, we'll both be looking for other posts,' said Susan calmly, gathering up her knitting and rising to her feet. 'Now, please excuse me, it is time for Falina's English lesson.'

Luiz came out to say there was a 'phone call for *Senhorita* Oliveira from Beira. He looked old and weary, she thought compassionately as she walked across the hall. Pepita's voice was plain to hear. 'It is nice here, certainly, Alfonso, but I did so enjoy my stay in Beira. What a gay city you showed me. Here we are so quiet, nothing ever happens,' she wailed.

As Susan threw her knitting on to the bed, she wondered who Alfonso was. Presumably another young man whose home Pepita and her mother had visited.

Susan woke the next morning tense and unhappy. This was no place for her, but she still had not found out a thing about Peter's death and wondered who else she could contact. Perhaps some men from the construction camp were on holiday? In that case Mark would not have been able to ask them about that night in Marula. It was not much of an idea, but it did lighten her thoughts. She quickly rang Mark and he was able to tell her that one man had just returned from sick leave. She hung on while he went to talk to him and came back to say that all the man could tell him was that he had seen Luiz in that area that night. Luiz?

Susan rang quickly down the steps and along to the patio where the old man liked to sit. His head had fallen forward in a doze, but raised himself stiffly at her approach.

'Luiz,' she said gently, trying to stifle the urgency that had filled her whole being. 'You know why I came to this country, don't you?'

He nodded and she went on, 'A man from the dam site has just told a friend of mine that he saw you running away from the scene of that fight in Marula. Oh, Luiz, please tell me about it,' she pleaded.

The dark eyes met hers unflinchingly but his face was troubled.

'I cannot, *senhorita*,' he said contritely. 'It has worried me much.'

A sudden sickening fear made her glance at him sharply. Had he been involved? No, those old eyes were much too honest.

'Did the *senhor* tell you anything else?' he asked with an anxious quiver and when she shook her head, he sighed as if a great load had been lifted from his stooping shoulders. '*Senhorita*, I cannot tell you more.' Although there was regret in his voice, there was a firmness too that was surprising.

'But Luiz, you do know something, don't you?'

A nod from the grizzled head. 'Won't you tell me?' she pleaded again.

'*Senhorita*,' looking at her with affectionate sympathy, 'I know the load you are carrying, but it is not my place to tell you. Others are involved and I owe them my loyalty.' He put out a trembling hand for understanding. 'Your brother is dead, but there are the living who must be spared. Please understand, *senhorita*.'

Bitter disappointment welled up inside Susan. She was so near the end of her quest and now it had been snatched from her. Had it been one of Luiz's own family who had been involved and then a more frightening thought. Someone from the *quinta*?

The old man watched her turn away, the troubled lines still on his wrinkled brow. He shook his head as he shuffled back into the house. It was better so.

# 8

Several evenings later, the *Marquês* held a dinner party to which Mark, Nannette and Nikki had been invited. They had had predinner drinks in the *sala* and Susan had been an amused onlooker when Pepita and her sister-in-law had been introduced. Their smiles of greeting were no more than a mere flicker of acknowledgment. Pepita looked especially charming in a deceptively simple white frock, that showed off her figure to perfection, while Nannette wore a flame silk sheath. She enquired where Pepita lived and what job she held.

Pepita's rich, warm laugh bubbled out.

'I wouldn't dream of working and I'd never willingly lose my dependence on our menfolk. It is so unfeminine to work. No, no!' and shuddered elegantly.

'But don't you get bored?' asked Nannette, offering a cigarette.

'How can that be, *senhora*, with so many clothes to buy and many social engagements to fulfil? I suppose you have to work?' She spoke pityingly. 'Now please excuse me, but I must speak to *Senhor* Fenn.'

159

Nannette silenced and biting hard on her upper lip, joined Susan. 'I'll stick to Alberto, thank you. I know when to get out of the running,' with another disgruntled look at Pepita as she lolled gracefully next to the *Marquês*.

Susan slipped away upstairs to take a peep at Falina and lingered on the balcony, loath to go back downstairs, but deciding it would be better if she did.

Nikki came and sat beside her. He had taken off his jacket as the night was warm. She had never seen him with his sleeves rolled up and her eyes flickered with surprise as they rested on a long scar that ran from his wrist to his elbow and the words of Carlotta came to her again. The man with the mark and something about great happiness. Susan also remembered the dark, forbidding look that had shuttered the face of the *Marquês* that morning. Well, the old *ama* was wrong, she thought with amusement, for she had no feelings at all for Nikki, except exasperation.

He had noticed her startled expression and proudly held up his arm. 'Mauled by a lion,' he grinned and entertained her with a graphic description of a safari he had been on. He was much put out when she excused herself.

The *Marquês* was the perfect host and not even Pepita, could keep him by her side for any longer than he deemed polite. He came over to Susan and without preamble said, 'You're very pale, *senhorita*. Are you worrying about this friend in England? *Senhora* Deane was telling me about him. Are you missing him so much?'

This took her by surprise. 'Felix?'

'Perhaps it would help if you told me about him,' he suggested quietly.

The *Marquês* in this mood was difficult to face. 'There's nothing to tell, *senhor*. I went out with him for a while, but when I started working in the evenings, he stopped seeing me,' and was unaware of the indignant note that had crept into the words.

'That was when you started to save for this trip?'

Susan nodded.

'So, that's why you've gone out with Mark Fenn, to try to forget this feeling of rejection? It's not a good thing to replace the man in one's affections with another so quickly, Miss Deane.'

This was too much for her.

'If you're trying to tell me that I might be caught on the rebound, *senhor*, you're very much mistaken. I never loved Felix and I don't love Mark Fenn either.'

'And yet you allow him to take you out?'

Susan was fast losing patience. 'Just

161

friendship can exist between the sexes — where I come from,' she added pointedly.

The *Marquês'* brows flared with disbelief. 'There is no such thing as a disinterested friendship between a man and a woman.' He shrugged. 'But then I can only speak from a man's angle,' he replied with a wicked twinkle. 'A Portuguese one. Hasn't Fenn made any gesture to you as a woman? If he hasn't, I haven't taken him for the man I thought he was.'

She lowered her eyes remembering Mark's light-hearted proposal.

'He really only took me out because he was lonely and to talk about Nannette,' and enjoyed the *Marquês'* start of surprise.

'He knew *Senhora* Deane before?'

She nodded. 'He was with Peter in Brazil and, for a while, so was Nannette.'

The *Marquês'* smile was sardonic. 'You do make things complicated for yourselves, you English,' he marvelled and excusing himself went across to sit beside *Senhora* Oliveira.

'What's his lordship playing at,' asked Mark with a grin as he sat down in the vacant chair next to Susan. 'Musical chairs?'

Susan chuckled. 'Don't you know the *Marquês* by now? I've never known a man to be quite so correct as he is.'

'Pepita doesn't care for it. I've had a most

enlightened talk with that young lady, about the restrictive life she leads. Believe it or not, Susan, she is envious of you.'

★ ★ ★

A few days later Mark rang and asked Susan to go out with him. He sounded tense and unhappy. She had heard through the grapevine that Alberto Lopez was staying at the *pousada* and that Nannette had been seen in his company.

'Mark, I'll go out with you on one condition that you bring me home at a reasonable hour,' and heard his exasperated sigh.

'All right, but ask his lordship if, as a special treat, you might have a pass until twelve. I'll pick you up at seven. Will that be too early?'

Susan assured him that it wasn't and ran upstairs to wash her hair.

As they drove along the road that evening, the last of the sunset pulsed and glowed, the peaks of purple mountains in the distance, were like ragged cutouts against the apricot and turquoise splendour of the sky and then swiftly the night placed a blanket over the countryside.

Mark had switched on his lights earlier and

every now and again some small animal scurried across their path to the safety of the tall grass on either side of the road.

'Mark,' she said at last, 'Isn't it better to get what is worrying you, off your chest?' unable to bear this heavy silence any longer.

'Nannette has given me my walking ticket,' he said baldly, throwing up a hand in a hopeless way, a gesture he must surely have learnt from the Portuguese.

'Oh Mark, I'm so sorry! Surely you didn't expect anything else. Not really?'

'I had had hopes.' The flat voice was devoid of all expression.

His suffering hurt her. 'Mark!' was all she could find to say.

They eventually pulled up at the entrance to the hotel, the music and ready laughter coming out to greet them as they made their way to a table in the grounds. Mark seemed disinclined to dance and sat idly twirling his glass, moodily watching its contents glisten from the string of lights that ornamented the garden, a silent, morose, chain-smoking man and Susan became more and more apprehensive as the evening wore on. Mark was drinking heavily by now and she was just about to suggest they leave when he said slowly, his speech thick.

'Why don't I just fall in love with you,

Susan. You are serene an' *sympatico*, no tricks up your sleeve. I'm very fond of you, Susan, don't you know? What about it?'

'Fondness isn't enough for either you or me, Mark,' and could only sit and watch him, as a mother might do for an ailing child, her compassion a silent outgoing of the spirit. 'No, Mark, you would always see Nannette when you took me in your arms,' and I would think of Raynor, she could have added.

Mark viewed her owlishly. 'Too bad!' his hand groping for hers as it lay on her lap. 'You wouldn't mind roughing it; a good bush wife.'

Susan glanced pointedly at her watch saying casually that it was time they were making a move and suggested that she drive.

'Certainly not,' he retorted, but made no demur at what he considered an early hour.

She watched him uneasily as he fiddled with his keys and was very relieved when he asked her, sheepishly, if she would mind driving and before she had driven out of the town, Mark was asleep beside her.

The journey back to the *quinta*, for Susan, was a nightmare and it felt as if she was crawling over the potholes. In this unfamiliar car she dared not increase her speed and the vegetation that had seemed so friendly, now loomed up menacingly ahead.

165

Susan began to worry about Mark. He wouldn't be able to drive himself home once he had dropped her and decided all she could do was to rely on the *Marquês*. He would know what to do, even though she knew he would be furious. Fingers of strain had crept across her eyes by the time she turned into the avenue and switched off the engine in front of the house. Mark was still sound asleep, his eyes half closed, his mouth open, emitting a gentle snore.

The *Marquês* met Susan on the steps and raised his eyebrows as he looked beyond her to the car. 'Isn't Fenn coming in?'

'He's asleep,' she said unsteadily. 'I — I didn't know what else to do, *senhor*, so I drove him here.'

His manner became instantly withdrawn, a nerve twitching at the corner of his mouth as he held back his anger from her.

'So, he has had too much wine, the *Senhor* Fenn?' his eyes expressing unutterable scorn.

'I — he — Nannette has told him she does not want to see him again. He — he loves her and it hurts,' stammering in her effort to make him understand.

'So now he wants you, so that he has a shoulder to cry on?'

'It's not very pleasant to find that one has made a mistake, *senhor*,' she replied wearily,

166

trying again to defend Mark.

The *Marquês* led her into the study and she nervously settled herself in a chair, longing to be dismissed so that she could go to bed.

'What are you going to do about Mark? We can't leave him there.'

'He can stay there all night for all I care,' he replied with angry intensity.

Susan's head came up quickly and gazed steadily across at the *Marquês*. '*Senhor*, can't you forgive one lapse? I can assure you that this has never happened before.' Her eyes dropped to her clenched hands. 'We all react differently when we've been knocked sideways.'

'I'm not at all interested in Fenn's emotional state, but what I won't tolerate is the way he's tearing you apart.' His voice softened. 'This pain will pass, *pequena*.'

'But I don't love Mark. Why can't you accept that?' she asked with exasperation. She looked very young and vunerable sitting there in her blue, black sprigged frock, dark smudges of fatigue under her soft blue eyes.

The *Marquês* was standing at the window, gazing out across the night filled garden, then turned to her again. 'Are you protesting because this infatuation is against your better judgement? This — this feeling you have for

Fenn, it can't be love,' he protested explosively. 'He's so careless of your wellbeing. *Senhorita*, it's the woman who should do the leaning, not the other way around.' There was a tiny pause before he continued. 'You, with your independent ways may smile at it, what you call, our old fashioned ideas and talk of the equality of the sexes, but where there is love, truly shared, surely it is a joy to a woman to be possessed completely in a loving partnership?'

Susan could sense, only too well, to her deep hurt, how this man would love and cherish the woman he made his wife and suddenly, to her horror, the tears spilled over and ran down her cheeks. She stumbled to her feet, driven by a desperate need to escape and then felt a hard hand on the back of her head and was drawn swiftly against him.

She stood still in the shelter of his arms, trembling violently. The evening had been too much for her.

'I did not mean to make you weep, little one. I won't leave Fenn out there and you're not to worry further, but first, I'll give you something to steady those nerves. That drive could not have been easy.'

Susan sat down again as the *Marquês* went across to the wall cabinet and came back with a glass of brandy. She protested, but on his insistence emptied the glass, pulling a face as she did so, the fiery liquid making her choke, but it calmed her.

'You were wise, Miss Deane and now you'll go to bed. Matters are usually viewed with a calmer mind in the morning.'

He had turned and as the light fell on his face, she became acutely aware of the tired lines sharply etched on either side of his mouth.

'You look tired, senhor. I'm sorry to have kept you up.'

She had suddenly disarmed him and watched with surprise, the hard shell of his face break up, an unreadable expression creeping into his eyes, softening them to a velvety darkness.

'I think you work too hard,' she added shyly.

Instead of his head being flung back haughtily and being told to mind her own business, his tone was friendly. 'Miss Deane, you are the first person, for a long time, who has sensed that,' and gave a characteristic shrug. 'It's the man's part to do the worrying, but your concern for me, I find very sweet. Thank you.'

It seemed wrong, reflected Susan, that no one had thought the fidalgo was, after all, human and that he too might have moments of intense weariness and troubles of his own. He carried a great weight of responsibility, a lonely man on a lonely road, but then he wasn't an approachable person either, as she herself had found out. Would his marriage make him any different?

'I am a little weary,' he conceded, 'but I leave for Portugal soon and the sea journey will give me the rest I need.'

Final arrangements for his marriage would probably be made, whilst he was overseas, Susan reasoned. Pepita and her mother would be leaving soon too. Would they all make the sea journey together?

'Miss Deane, have you had anything to eat tonight?'

'There were bits and pieces, but I was too worried about Mark and how I was to get back — '

The *Marquês* exploded, but it was a little boy crossness, not the white hot anger she had seen him display on occasions.

'And how do you think you can exist on that until morning? Come, we'll go and see what we can find in the kitchen.'

Most of the house was in darkness as they went across the lawn, the *Marquês* suggesting that she wait in the courtyard and returned a few minutes later with a glass of milk and a plate of homemade biscuits.

Susan's heart thumped over. What surprising things he did. She would always remember them, even after Mozambique had become just a dream. She would remember the way his hair grew thick and dark from a wide forehead; the shape of his hands and the deep chuckle that was sometimes surprised out of him and yes, even the flame of anger that so often sprang up between them. That anger that could be so devastating and yet, the woman he loved would be warmed and protected by those self same fires that burned within him.

As Susan finished her milk and ate her biscuits, her glance went, deliberately, around the courtyard. 'Please tell me the history of this old property.'

'Would you like to go up on the roof and see why this house has this unusual layout?'

She quickly rose to her feet. Anything rather than having to endure sitting so close to this man. His nearness was suffocating her.

The *Marquês* unlocked the door that revealed a steep flight of stairs, his hand beneath her arm, strong and warm.

'We keep this door locked. It would not be safe for Falina to wander up here.'

They came out on to the roof, the surface roughcast. 'Nothing has changed up here,' he explained. 'This is how it was when the house was used as a fort,' and showed her, tucked away in a corner under a tin shelter, two small cannons, then led her to the wall that ran around the entire building.

Susan stood looking out between the battlements and could make out the stables close by, but the surrounding countryside was blanketed in darkness. She was very aware of the *Marquês* close behind her, his warmth permeating through her thin frock, as he pointed out the winking lights of a distant town.

'All this stone work is interesting to us today,' he said, moving a little away, 'but in the days of my great-grandfather this was where men fought for the right to keep the land they had cleared from dense bush, some of it primeval forest and make no mistake Miss Deane, it was just that.'

His voice found an answering thrill through her as he spoke about the past.

'Before my grandfather started to renovate this place, there were no windows on these outside walls and only one door which is now the main entrance. When that was shut, no intruder could penetrate. The storerooms were kept continually stocked and even sheep and goats lived in the courtyard in times of trouble. It has been, all my life, a proud heritage that I will one day, pass on to my son.'

'Have you thought of writing the history of the *quinta, senhor?*' she asked. That Pepita should bear him a son, was a tormenting thought.

The *Marquês* shrugged and said with regret, 'Perhaps when I'm an old man? Of course records have been kept.'

'Do you think I could sort them out for you?' she asked eagerly.

'They're in Portuguese,' smiling down at her disappointed face. 'But when you have become proficient in our language, perhaps this is what you could do for me?'

Susan turned away. He was teasing her, for her stay in Mozambique was drawing to a close and the thought brought a sharp pain.

They had just come back into the hall when Pepita came through the library door, holding a book against her soft white dressing gown of pleated nylon with much lace at throat and wrists, her hair hanging down over one shoulder in an enormous plait. Her start of surprise turned to one of disapproval and Susan felt the *Marquês* stiffen.

Perhaps he was already regretting the moments he had spent with her. With a hasty 'goodnight' she fled upstairs.

Nannette phoned just as Susan was

finishing her breakfast to tell her that she was leaving for Beira that afternoon. Susan hurriedly drove into Marula and found her sister-in-law throwing her clothes into a big, expensive suitcase that lay on the bed.

'Cold, arrogant, proud — FISH!' she muttered. 'I don't know how you can work for that man.'

Susan looked startled. 'What on earth has he done for you to be so angry?'

'I asked him to come and see me last night about Peter and nearly made a fool of myself.' In went a finely pleated nylon skirt, followed by a pile of panties.

'Nothing wrong in that, surely?'

'Most men of my acquaintance would have made the most of the opportunity, but no, not his lordship. I thought all Portuguese men could never resist an attractive woman,' throwing another garment into the case. 'I'm not used to being given a brushoff like — like,' looking around for a simile, 'an ant on a wall.'

'You mean he wouldn't fall for your charms?' Susan took a dress from Nannette's clutch, proceeding to pack methodically.

Nannette lit a cigarette with hands that shook. 'Well, he did confirm what you had already told me, that he was engaged, only he put it in rather a funny way. That he was affianced in his heart and matters such as this could not be hurried. Oh, Susan, I don't know what has happened to me.' She almost hissed the words. 'I must be losing my touch — getting old, or something,' drawing furiously on her cigarette again.

Susan glanced up sharply, not liking the cold, hard expression on that lovely face. Nannette in this mood was liable to do anything and was not really surprised when she said, morosely, 'I'll marry Alberto, that's what I'll do. Oh, yes, I know what you're going to say,' she went on hurriedly. 'What about love? I loved Peter and look what happened to him.' Her voice faltered and a haunted expression crept into those deep blue eyes and then hardened again. 'At least Alberto will adore me and I'll be able to twist him around my little finger. What's more, he'll give me all I want.'

Susan was shocked. 'Nannette! What's got into you! You used not to be like this!' and turned away, appalled by her attitude.

'It's a cold, hard world,' she retorted fiercely, 'or have you forgotten, with all the soft living you've had lately?'

'At least I've kept my self-respect,' flashed Susan.

'Don't become my conscience, please. I just couldn't stand it.'

'And what about Mark?' she couldn't help asking, busy folding a frock in tissue paper.

'What about him?' hedged Nannette defiantly.

'Just that he loves you.'

'There you go again,' giving an impatient exclamation. 'If you think I'm going to tie myself again to a man with dam construction in his veins, you've got another think coming.'

'He did say he'd settle down in an office job.'

'And I'd have to live with a man with a permanent grouse. Alberto is fabulously wealthy, with property here and in Portugal,' she enthused, eyes overbright. 'You can keep your *Marquês*. He has completely ignored me ever since I arrived.'

'He's not mine,' Susan reminded her, keeping her voice steady.

'No, I shouldn't think so, either,' she replied nastily with a pointed look at the clear, tanned skin, innocent of make-up. 'That man has a nice taste in women, judging by Pepita.' Her words had taken on a vemonous twist.

They were interrupted by a knock on the door. It was one of the waiters to say that *Senhor* Lopez was waiting in the small *sala*. They went downstairs, Nannette gay as if she hadn't a care in the world and introduced

Susan to Alberto, flinging back her dark hair over one shoulder, but the fat little man had eyes only for Nannette, patting her hand and saying that they had to leave at once. When the suitcases were brought down, Susan hurriedly gathered up her hat and took leave of them.

★ ★ ★

Luiz, as usual, came out to meet her and was conscious of the troubled look he gave her and hoped the old man would feel that he could confide in her. Luiz though, was not her only problem. Falina, of late, had become withdrawn and not her usual happy self. She arrived back from school half an hour later and Susan taxed her with this and the flood broke.

'I hate Pepita! I don't want her for a sister-in-law!'

'I quite agree,' replied Leonora who had come in at that moment. 'You know, Susan, she boasted the other day of having boyfriends here and in Portugal. I suppose she had to say that in case Raynor doesn't offer for her.'

Susan's heart leapt. So no definite plans had yet been made. 'Leonora, you shouldn't gossip.'

'I think we should discuss Pepita for — for if she marries Raynor, she'll live here, very close to us.'

'I don't want her to marry Raynor,' said Falina tearfully. 'I heard her say to her mother that I was going to be a nuisance and other arrangements would have to be made.' Her eyes were big with apprehension.

'And what did you do to make her say that?' asked Leonora accusingly.

'She smacked me hard, because Kitten got into her room. Pepita doesn't like cats,' Falina remarked with surprised indignation. Anyone not liking cats was beyond her comprehension.

'What mischief was that demon up to?' asked Susan resignedly.

The little girl gave a watery grin, showing a gaping hole where the day before a tooth had been.

'He only sharpened his claws on her suitcase. You know cats must sharpen their claws?'

'But not on someone's expensive suitcase. Falina, you must keep that kitten under better control.'

'Would Pepita send me away, Susan?'

Susan caught her breath, but managed to say calmly, 'Of course not, darling, this is your home.'

'Don't be stupid, Falina,' said Leonora impatiently. 'Raynor would never consent to that. If Pepita thinks she will be able to curl him around her little finger, she's in for a big shock.'

When she climbed into bed that night, Susan, with a worried frown, wondered about Falina's fate if the *Marquês* should marry Pepita.

\* \* \*

The next day passed in a haze for Susan as she helped the old *senhora* take out all the priceless treasures from the glass-fronted cabinets in the *sala* and diningroom. Each piece had to be carefully cleaned, a chore that could not be left to the servants.

She had never seen *Tia* Maria so happy as Leonora's eighteenth birthday drew near. It would be a great affair, families from all over Mozambique attending. Those who had a distance to travel would stay overnight at the *quinta* and all the rooms at the *pousada* had also been booked for that Saturday night. Not that there would be much of the night left, after the ball, but guests would want a few hour's sleep before starting on their return journey and would the *senorita* please do the flowers as usual?

All this and more the old *senhora* poured into Susan's ears as they went from one cabinet to the next, but all the cleaning had to be set aside, when *Tia* Maria came back from a 'phone call, her face crumpled with worry, her black bead eyes sunk deep in her head.

'Miss Deane,' she panted, 'I've been called away to visit my sister. She's very ill and what will *Senhora* Oliveira and Pepita think, I just do not know.'

Susan suddenly felt sorry for this dumpy little woman and gently took her hands in hers, stilling their restless worrying.

'I'm sorry, *senhora*. Is there anything I can do?'

'Would you see that the servants do what they are supposed to do? All the meals have been ordered for today, but they do need supervision. I hope to be back after breakfast tomorrow,' but the *senhora* was not back, Susan receiving an agitated call the following morning, the old voice shrill with distress as she went on to say that she would not be back that day.

Susan scribbled down the menu and told *Tia* Maria not to worry.

'But *senhorita*,' came a wail, 'I have not told you everything. The *Marquês* is expecting four men to dinner. There's an important meeting afterwards.'

180

'I'm sure Timbo will manage,' Susan assured her.

It was the first time she had heard *Tia Maria* so conciliatory. Her profuse thanks poured over the 'phone.

Susan took the menu to the kitchen and supervised the lunch, but the blow fell later that afternoon, as she was giving Falina a lesson. Luiz had sought her out and with an anxious frown told her Timbo was ill.

'It just had to happen when the *senhora* is away,' he groaned. 'I've been to see him and he's shaking like a leaf. It must be the malaria. What about dinner tonight, *senhorita?* Would you like me to hire a cook?'

She shook her head. 'I'll cook the dinner. The *senhora* forgot to give me a menu so I'll think up something myself.'

The old butler's hands opened in a shaky protest, the rheumy eyes startled. 'I'm sure the *fidalgo* would not expect that.'

'Don't tell him, please Luiz. The meeting after dinner is an important one. I don't want him worried,' but her mischievous eyes did comfort him, somewhat.

Susan and an excited Falina went down to the kitchen a little later to find Salvador already there, making up the fire. Susan eyed the black monster nervously and wondered what she had let herself in for. A coal stove was an unknown quantity, but found it easier that she had expected, for Salvador knew just how to control the all important damper.

She decided on a simple meal consisting of vegetable soup, roast chicken, chipped potatoes, vegetables and a fruit salad.

'I'll also make my favourite cream cheese dessert,' she said to Falina, who was thoroughly enjoying herself.

Sometime during all the hustle and bustle, Luiz poked his grizzled head around the door. 'The guests have arrived and are having drinks in the *sala*,' his face long and lugubrious, 'and the *Marquês* has asked that you join him for dinner, *senhorita*.'

Her eyes widened in dismay. 'You'll just have to ask him to excuse me. You haven't told him that Timbo is ill, have you?' her voice anxious.

'No, *senhorita*, but now he has especially asked you — what do I say?' shaking his head uncertainly.

'You must not tell him I'm here in the kitchen. This is a most important business dinner, the *senhora* told me this morning and I can't leave Salvador here on his own. I only hope that they all like my dinner.'

'I do not know what the *fidalgo* is going to say,' muttered Luiz, with worried eyes as he shuffled out of the kitchen.

Everything went smoothly. Salvador served the last course and came back a beam splitting his face. 'Everyone, *senhorita*, has ordered a second helping of your dessert.' He was as proud as if he had made it himself. 'One of the *senhors* said to the *fidalgo* that the dinner was very good. Timbo will be plenty finished when he hears that. And now, I go and do the washing up.'

Susan and Falina left him to his chores and went to the cloakroom to wash their hands.

The mirror showed Susan a most unflattering reflection. Hair ruffled and damp from the heat of the kitchen, a nose that shone like a beacon and her once crisp cotton dress limp about her slim body.

'Oh well, what does it matter,' she said to Falina, combing her hair, while that little girl retorted that they had had a lovely evening and hoped that this could happen again.

Susan chuckled as they made their way to the seat in the courtyard, where it was blessedly cool.

'We'll sit here for five minutes and then it's bed for you, my poppet.'

The perfume of the flowers bordering the lawn was so much stronger now than in the heat of the day. The kitten had joined them and was playing some game of its own amongst the mysterious shadows cast by the shrubs.

They were laughing at the kitten's antics, when the *Marquês* strode over to them and Susan saw that he was at his most forbidding, disapproval in very line of his body, a stony brilliance to his dark eyes.

'So, it was for no reason at all, that you refused my invitation,' he said tight-lipped.

Susan had stood up at his approach and was about to explain when Falina confronted him like a small, angry tornado.

'It was Susan who cooked your dinner and all you can do is be horrid.'

Susan was so startled at this unusual behaviour that she could only stare down at the little girl and even the *Marquês* was taken aback, but the storm had outridden itself and Falina threw herself against Susan, sobbing noisily.

Pepita had joined them, her whole manner one of shocked incredulity.

'What a to do!' said Susan calmly, her arm around the still sobbing child. 'Go and bathe your eyes, I'm coming up soon.'

'I've never come across such atrocious manners. That's the trouble when you admit a foreigner into your household,' Pepita tittered angrily.

'I think it might be a good idea if you go

and make your peace with Falina,' said Susan matter-of-factly to the *Marquês*. 'This has been nobody's fault.'

He had never been dismissed by a woman before and even Pepita was a little shattered.

'I'll see you, Miss Deane, after my guests have gone,' he said coolly, signalling courteously to Pepita that they go back to the *sala*.

'What all the fuss is about because a governess won't dine with you, I just can't understand,' Susan heard her say as they walked away. 'I certainly won't have an English Miss in my house when I marry and why wasn't that child in bed hours ago?'

How right they looked together, thought Susan, with a sharp pang of pain. All the evening's pleasure suddenly evaporated. Her senses quivered as she went upstairs and heard the murmur of voices from the *sala*. The guests partaking of after dinner drinks before the business meeting commenced and Susan had no doubt of Pepita's ability to play her part perfectly.

Falina was still tearful when Susan tucked her beneath the mosquito net, but hoped that with all the excitement, she would soon drop off to sleep. Leonora came in a little later to congratulate Susan on an excellent dinner.

'Pepita was upset when the gentlemen asked that you be especially thanked. That one likes all the praise,' adding, 'I'm going to bed,' stifling a huge yawn.

'Leonora, you can't do that! What about *Senhora* Oliveira and Pepita? Who's to entertain them?'

'It — it was at their suggestion we have an early night. Pepita knew she won't see Raynor again. The meeting could go on for hours.'Saying 'goodnight' she went to her room.

Susan knew she would have to wait up for the *Marquês*, no matter how long the meeting lasted. Suddenly she had a longing for home and all the safe relationships that had been hers.

Two hours later the *Marquês* came into the upstairs lounge and Falina immediately called out to him. He raised an enquiring eyebrow.

'She's very restless, *senhor*. Please go and tell her you aren't cross with her.'

Falina hurled herself into her brother's arms, a small, tearful figure in shortie pyjamas and tousled head.

'I'm sorry I was rude, but I had to stand up for Susan,' she hiccoughed.

'Yes, of course, *pequena*,' he soothed. 'We must always be loyal to our friends and now, you must go to sleep. It's very late. I'm not angry with you, Falina. I did not know that our Miss Deane, with your help, cooked our dinner,'stroking the dark hair from off her hot, damp forehead.

'Are you sure?' she asked anxiously.

He nodded and Susan was amazed at his gentleness with his little sister.

'And you're not angry with Susan, any more?'

'No,not with her, either. Now come, Miss Deane will tuck you up again.'

Susan had been remaking the crumpled bed and as she bent down to kiss the little girl before tucking in the net, Falina said drowsily. 'Ooh, you do smell nice.' All was right with her world again.

'Only soap and powder, funny face.'

'I don't like it when Pepita hugs me. I want to lose my breath.'

'Not only your breath do you lose, but also your head,'murmured the *Marquês*, with a wicked glance at Susan as they went into the lounge.

He waited until she was seated, then said abruptly, 'Once again I must apologise, *senhorita* and my thanks for what you did this evening. I could easily have taken my guests to the *pousada*.'

'The *senhora* impressed upon me that it was an important business dinner, *senhor* and I did not want you to be worried about such a small matter. I enjoy cooking,' then added quietly, 'but I do understand that you must have thought me very rude when I refused your invitation.'

He was sitting relaxed, the shadows from the table lamp etching hollows under his high cheek bones.

'You must always put others first, must you not, my little English *menina*,' and smiled before going on. 'I suppose you will now say that this goes to show that Leonora should know something about cooking?'

'I think it would be a great help to her,' she replied, peeping at him from under long lashes, an action that would have done Nannette or Pepita credit, but she was unaware of her mischievous eyes.

Previously *Tia* Maria and the *Marquês* had been against Leonora learning to cook, because she would always have adequate staff.

The *Marquês* gave a low laugh. 'It is quite amazing how circumstances have a habit of working out for you and my authority goes for nil. How can I possibly say now that there is no need for my sisters to learn to cook, when I have seen my guests forget to argue as they ate that excellent dessert. A little port mellowed them further. We got through the meeting in a very short time.'

# 9

The night of Leonora's party had arrived. The *quinta* blazed with lights and in the great stone-flagged entrance hall the *Marquês*, immaculate in formal dress, awaited his guests.

Susan, with a deep satisfied sense of achievement, watched Leonora standing between her brother and Enrico. She bubbled with happiness that showed in her hair, in her face and in her sparkling eyes. She was beautiful standing there in a cream lace frock over a pale turquoise petticoat, especially ordered from Portugal. Glowing emeralds around her neck and in her piled up hair, were a gift from her *noivo*.

Susan's glance moved around the hall, where earlier, she had banked masses of pointsettias and had mingled red-hot pokers with chaste arum lillies against the wide staircase as it rose majestically to the upper floor.

For the ballroom she had chosen varying shades of pink gladioli and the feathery mauve of the ginger bush. Now the guests were walking up the shallow steps to the door.

The *Marquês* bent over the hands of the ladies and had a warm smile for the men.

Susan had been pleased with her dress in her favourite shade of blue, the softly draped bodice and very full skirt, suited her slim figure, her only ornament a string of pearls. The frock had been expensive, but worth every penny, or so she thought, until she saw Pepita. Who would spare her a second glance when the Portuguese girl stood there in a sheath of flame brocade, peeping beneath a gold lamé evening coat that later she would, no doubt, discard.

She was standing now a little behind the *Marquês*, with a charming air of effacement, for after all, this was Leonora's evening, her whole attitude seemed to say; her lovely face serene, the only artificiality, the scarlet of her lips and the upward markings of an eyebrow pencil at the corners of her dark eyes that lent her an oriental look.

The orchestra, especially brought up from Beira for the occasion, started with a lively Portuguese number and Enrico led Leonora on to the floor, amid much clapping from the guests sitting on the stiff-backed chairs, with their dark blue damask coverings. The great doors of the ballroom were open, the heavy velvet curtains moving in the slight breeze. No doubt couples would disappear discreetly through those doors, if they could avoid the watchful eyes of the *duennas*.

The polished oak of the heavy furniture shone under the soft beauty of the chandeliers, which, Susan noted, had not been wired for electricity and which were so much kinder to the olive complexions of the *senhoras*. Jewels glowed, faces smiled and a hum of excited chatter went around this huge room as the *Marquês* led out *Senhora* Oliveira and then it was Pepita's turn, her head held up regally. All eyes were on her and she knew it, as if she were already mistress of this *quinta* and of the *Marquês*' heart. The guests gave each other knowing nods and wondered when they would be told of yet another betrothal in the House of Monteiro.

Susan watched this couple too. How beautifully they danced together, the *Marquês*, in a light-weight evening jacket supurbly cut was, she had to admit, a handsome man, when one became used to the high cheek bones and the dark inscrutable eyes.

She was talking to *Senhor* Cruz, when she was startled by a hand on her shoulder. It was Mark.

'I didn't know you had been invited,' her glowing eyes showing her pleasure and introduced the two men.

*Senhor* Cruz waved them away with a fatherly smile and as they joined the dancers, Mark said, 'It was practically a royal command that reached me.'

'It was kind of the *Marquês*,' she replied happily. 'It is wearying trying to chat to all these people. Not many speak English and although my Portuguese has improved, I still find it difficult to speak naturally. Not that I presume to speak to many, but there are those I do know.'

Mark twirled her down the long room, her laughing face raised to his. They had just passed Leonora and Enrico when she met the burning eyes of Nikki and hurriedly turned away, but that glance had sobered her. He would, no doubt, ask her to dance later on in the evening. They passed the *Marquês* and Pepita, her body pressed close to his.

It was a strange, jumbled up world, reflected Susan ruefully. Nikki wanted to dance with her, while, though she feared the close contact of her employer, longed to be in his arms and Mark's thoughts were, probably, not totally here.

She had sensed his withdrawal as soon as they had begun to dance.

It was only after supper that the *Marquês* claimed her for a dance. He had, as usual, been most correct all evening and had danced, after that first dance with *Senhora* Oliveira and then with Pepita, with all the ladies, in what appeared to be, a strict order of seniority.

No eyebrows were raised as he led her on to the floor. The *fidalgo* was merely doing his duty, although it did seem strange that the companion of his sisters should be asked to dance at all. A very nice girl, you understand — expressive hands were fluttered — quite devoted to Falina and what a difference in Leonora, so the old *senhoras* said to each other and then forgot about Miss Deane as they went on to speak of other, more important subjects.

Susan had stiffened when the *Marquês* had first put his arm around her, but he was such a good dancer that she found herself relaxing.

'Ah, that's better, Miss Deane,' he murmured, but not another word did he utter and she was content to let it be so.

She could feel the warmth of his hand as he held her lightly and when she did raise her head, found him looking down at her with an unreadable expression on his lean face. Had his arm tightened around her just a little, or had she imagined it? A dreamlike quality stole over her, realising she would never be given another opportunity such as this. A roll of drums heralded the end of the foxtrot. The

*Marquês* took her back to her seat next to *Senhor* Cruz. With a tiny bow he was gone and Susan saw him lead Pepita out again.

Nikki was her next partner, gypsy dark and bold, allowing him to swing her into a waltz, with a certain amount of misgivings.

'My apologies, Susan,' he said softly into her hair, his arms tightening around her. 'If I'd had my way, I would have led you out for the first dance, but no!' His voice was bitter. 'I had to conform. All those *senhoras*! Hé!'

Her lips quivered with amusement. 'I did see you with some beautiful girls too, Nikki. Surely they made up for all the *senhoras*?'

'But not one would come out with me into the garden,' he mourned.

She laughed. 'Perhaps their mamas had warned them about you?'

'What matter?' his reply was contemptuous, as his hot glance, reflecting his admiration, roved over each of her features until Susan longed to walk off the floor.

'You are so beautiful! You make me want to do something desperate,' he whispered hoarsely.

'Just because of my fair hair? No, it's Pepita who is beautiful. I have never seen such exquisitely dainty features and those eyes — '

'Forget Pepita,' he muttered. 'It is you I have, at last, in my arms. Susan!' holding her closer until she could hardly breathe. 'This is what I've longed to do ever since I first saw you.' He grinned wickedly down at her, enjoying her outrage, twin devils in those dark, intense eyes.

'Nikki, behave yourself! What will all those *senhoras* think?' she protested, trying again to push herself away, but his arm was like an iron band.

'If you do not struggle, nobody will notice,' he said softly.

He was unfortunately right, thought Susan furiously and was thankful when the waltz came to an end.

'I'll come and ask you to dance again.'

Rather than face that, Susan slipped upstairs and on to the small balcony that overlooked the dance floor. Here she found Falina, nodding sleepily, but determined to miss nothing.

'When is Raynor going to announce Leonora's betrothal?' she asked, leaning tiredly against Susan.

They waited a little while longer, watching the gay scene below and then saw the *Marquês* step on to the dais at the far end of the room, Leonora and Enrico joining him and the announcement was made, midst a burst of clapping.

'Wasn't Leonora beautiful, Susan. Now she

can be happy and not lie on the bed, doing nothing, any more. I'm — I'm going to miss her,' with a watery sniff.

'She won't be getting married too soon. Come along to bed, darling.'

* * *

All the guests slept late the following morning, the dancing having gone on into the early hours. Susan had heard some of the cars leave, but had quickly dropped off to sleep again. Now she slipped out into the dew-filled garden and nodded to the horde of servants who were already polishing the huge ballroom floor.

Luiz met her and greeted her warmly. 'You are up early, senhorita. I will see about some breakfast for you.' He was just about to turn away, when Susan laid a hand on his shoulder.

'Thank you. Luiz won't you tell me what you know about that fight in Marula? At least whether it was my brother's fault, or not? That's all I want to know.'

The troubled look was back in the old man's eyes, but this time, she was quick to sense his indecision.

'I promise, whatever you tell me, will go no further.'

'I know nothing,' he muttered but as she turned to go, he said hastily, as if the words were forced out of him. '*Senhorita*, I have much to thank you for. You saved the mother of baby Susan, who is a distant relative of mine. I have not presumed to tell you this before.'

There was another painful pause and she held her breath. Would Luiz tell her at last what she so longed to hear?

He said brokenly as if struggling with some strong emotion. 'All I saw was Nikki rushing away, dishevelled, his coat torn and eyes so wild that I do not think he even saw me, then a man fell to the ground. That is all I noticed before I went after Nikki.'

'Who else was there, Luiz?'

The old man looked bewildered. 'I think there was another man, but I had eyes and thoughts only for Nikki. That a Monteiro could be in a drunken brawl — it was terrible! I did not know then that it was *Senhor* Deane I saw on the ground.'

Susan took his trembling hands and held them comfortingly.

'This is why you have been afraid to say anything, in case the *Marquês* should hear about it?' Now she understood why Luiz had been loath to tell her what happened that fateful night in Marula. The proud name of

197

Monteiro must not be besmirched and so the old butler, in his loyalty to the *fidalgo*, had borne this burden. It said a lot for the love and respect all the people had for the *Marquês*, that this had not reached his ears.

Luiz was grateful for her ready understanding. 'The *fidalgo*, he — he is the one I most care for and it would break his heart this — this scandal. Nikki — ' He could not go on.

'I understand. May I speak to Nikki about this?'

'Perhaps he'll tell you more. I don't know. Since that young man has left school I feel that I don't know him at all,' he said sadly, with a shake of his head.

She would ring Nikki, decided Susan, picking up her spoon and starting on the grapefruit that Timbo set before her.

Nikki was very surprised when he received her call and promised to meet her at the stables the following evening. He put down the receiver before Susan could protest and as she glared angrily at the 'phone knew, without a doubt, that Nikki was congratulating himself at the coming assignation. How despicable of him!

★　★　★

It was while the family were at breakfast the following morning that a cable was rung through. The *Marquês*, fortunately, took the message and was able to deliver it in a calm voice.

'It is for you, *Senhora* Oliveira. Your husband is ill and would like you to return home immediately.'

Pepita burst into tears and *Tia* Maria started to moan softly.

'When will we be able to leave?' asked *Senhora* Oliveira quietly, only her eyes reflecting her inner turmoil. She should not have come out here, knowing that her husband was not well, but who else could have brought Pepita on this delicate mission of the heart?

The *Marquês* rang through to Beira and was able to get two seats on the late afternoon flight.

'Oh, Raynor,' sighed Pepita sadly as she rose from the table, laying a hand on his arm. 'I was so looking forward to our boat trip to Portugal and now it is not to be. Poor Father, but then he's always ailing.' She turned to her mother eagerly. 'Is it possible that Raynor could ring now and see how Father really is?'

'We must leave on this evening's flight. If something should happen to your father, I'd never forgive myself.'

'Then, is it not possible that I might stay here?' she asked pitifully.

Susan caught the unexpected shaft of fear and uncertainity in her dark eyes, that were so deeply fringed.

'Yes, please let her stay, *Senhora* Oliveira,' *Tia* Maria said quickly. 'It does seem a shame that the poor child's holiday should be cut short, so abruptly. We would look after her very carefully. Don't you agree, Raynor?' turning to the *Marquês* for affirmation, but his manner, though courteous, was distant. 'It is better that Pepita go home with her mother.'

*Senhora* Oliveira's reply was sharper. 'You know, as well as I do, *Senhora* Ferreira, that it is impossible to do what you have just suggested and now you will come and help me pack, Pepita.' It was a command.

The girl's shrug was eloquent as she followed her mother from the room, an angry flounce to her step.

The visitors left after morning tea, *Tia* Maria openly weeping as she stood waving from the steps.

'Hé!' she groaned, as the car disappeared down the avenue. 'Everything goes wrong.'

'I'm glad Pepita's gone. She hated my cat,' said Falina with a pleased smile, which called down a tearful rebuke from the old *senhora*.

That evening Susan hurried down the flagged path to the stables, a little shiver of apprehension running through her. She was very much against this meeting and what was Nikki playing at, anyway? Surely he could have come to the house and they could have talked there?

It was a dark night and she stumbled on the uneven paving, the moon hidden by a deep bank of clouds. It was only her burning desire to get to the bottom of this matter of her brother that drove her on. She walked along the wide verandah, where there were a couple of dim lights, she was thankful to see, past open horse boxes, conscious of the soft rustle of straw and the warm, sweet smell of well-housed animals, Nikki was standing just inside the fodder room.

'*Olá, senhorita,*' was his gay greeting. 'Isn't it convenient that Raynor had to go to Beira this morning?'

'How did you know?' she asked suspiciously.

'I have ways,' he chuckled.

How had he known so quickly, she wondered uneasily. It could only have been *Tia* Maria, who must have rung him. She would have preferred him to think that the *Marquês* was at home and knew a mounting sense of fear.

'Oh, Susan, I'm so happy that you wanted to see me,' and before she could sidestep, he had caught her to him.

She was conscious of the hot desire of his arms and jerked away with sickened fright, anger spilling over as she faced him.

'I shall scream if you touch me again,' she warned. 'And really, Nikki, couldn't you have come up to the house?'

He looked down at her, all velvety persuasion. 'I enjoy this kind of stolen meeting, don't you? It lends enchantment to the evening,' his hands waving expressively.

'Why?' she insisted.

Nikki's glance was sheepish for a moment, then shrugged an impatient shoulder. 'Leonora or Falina would have seen me and told Raynor. As you know, I've been warned to behave myself or else — ' snapping his fingers defiantly, 'back to Portugal for Nikki Monteiro, like some naughty school boy.' Bitterness seeped into his words. 'Why couldn't you have come earlier?' he demanded. 'I was nearly caught by Joao.'

'Falina was restless.'

Nikki was still close to her and tried to take her hand with a bold appraising look that she hated and it was then he asked her what she wanted to see him about.

'Is it to say that you, at last, want me to

take you out somewhere?' he asked eagerly.

Susan shook her head. 'Nikki, you know why I came out to Mozambique, don't you?'

'To find out what really happened to your brother,' he replied too promptly.

'Yes. Nikki, I want to know from you, what really happened, that night in Marula.'

He turned to stand directly in front of her, his dark brows meeting menacingly, his eyes as hard as the pods of the cassia tree that grew in the stable yard and even though there was a smile on his thin lips, Susan knew another stab of fear. She must have been mad to have consented to meet Nikki here.

'So, old Luiz has spilt the beans, has he?' he said savagely. 'It is time he was pensioned, that one! He's too old for the job.'

'He only told me that he had seen you running away from the scene of the fight. That old man is very loyal to the house of Monteiro, Nikki. You should know that and should be thankful that he did not report you, especially seeing that Peter lost his life. It was only after many times of asking that he told me about seeing you and gave me permission to ask you what really happened that night. Please tell me, Nikki,' she urged.

'I suppose he felt he had to tell you, because you saved his relation's life.' There was a trapped look about his glance. 'You promise not to tell Raynor?' and as she nodded, went on almost humbly. 'Your brother was a true gentleman.'

'Go on,' she prompted, after the silence had gone on far too long.

When he did speak, the words came out dramatically. 'He saved my life. Peter could see that I was getting the worst of the fight and, I suppose because of his friendship with Raynor, pushed me aside.' His voice saddened. 'He took the blow that had been meant for me. This — this someone had a heavy baton and if Peter hadn't pushed me aside, I would have taken that hit and been killed. Don't hate me too much, Susan,' he pleaded. 'It was an accident.'

Strangely, hatred was far from her and only a surge of relief filled her. She dropped her head into her hands as a fierce sense of gladness tore through her and tried to stop her shivering reaction. So Peter had gone to Nikki's aid and that was the extent of his envolvement.

'Are you all right?' he asked in alarm.

All she could do was nod her head mutely.

'Are you shocked, Susan, but I did, at least, manage to keep the knowledge from Raynor,' he added quickly, as if he should be congratulated on that score. 'My cousin's anger would have been too terrible to contemplate if he'd found out I was mixed up in that sort of affair. A blot on the ancient and well revered name of Monteiro.' A shudder shook his lean frame.

'But, you too bear that same name. What made you do it?'

He drew hard on his cigarette, then exhaled noisily. 'You do not understand, you cold English, what it is like when the blood runs hot at *fiesta* time, the crowds and some fellow cheated me at cards,' he added bitterly. 'It is easy to stir the Latin temperament. I'm sorry! Won't you let me take you out, just once to show that there is no ill-feeling?' He had become an eager, pleading boy and had drawn closer to her, gently touching her hand with uncertain fingers. 'This knowledge will be a bond between us.'

He was as resilient as a rubber ball, thought Susan, but she made it clear, once again, that she had no intention of ever going out with him.

'I'm very grateful though, Nikki, for telling me what really happened that night. It's a great load off my mind.'

They had stepped off the verandah that ran the whole length of the stables and turned in the direction of the house, when Susan turned around, her eyes dilating with horror.

'Nikki, the stables are on fire,' she

screamed running back, shadowy forms already rushing about the yard.

'You must have dropped a cigarette end Nikki, in the fodder room,' then gasped. 'The horses!' but Nikki wasn't there any more, Leonora was.

She had seen the glow from her window.

In an incredibly short time it seemed as if the whole block of stables sprouted flames and Susan realized, with a horrible jolt, that it was only wooden partitions that separated the animals. The wood was old and as dry as last year's grass.

Suddenly she was plunged into a nightmare of screaming horses, battering their boxes as she and Leonora rushed down the long verandah, throwing bolts to enable the frenzied animals to make their escape.

Susan did not know when she had become aware of the *Marquês* and did not have time to register the fact that he had not stayed in Beira to see *Senhora* Oliveira and Pepita off on their plane. It was only after all the horses had been released, that he came up to her as she sat on the water trough, gazing with horror filled eyes at the flames leaping skywards. There was no hope of quenching the fire for there was no adequate equipment to deal with it.

She had never seen the *Marquês* quite so

angry before as he stared down at her, seated in a dejected huddle.

'What I want to know is, why I find you and Nikki here. Joao said he had seen you both earlier on this evening. Why is it every time my back is turned, Nikki is at the *quinta*?'

'This time it is not his fault, *senhor*, but I cannot tell you the reason why I asked him to meet me here. I have promised,' she added quickly and suddenly felt sick. This was awful. The *Marquês* now thought she and Nikki — She shuddered.

'I'm very well aware that all my cousin's affairs are shrouded in secrecy.' His voice had taken on a bitter, ironic note and brushed his hand across his eyes as if to blot out something distasteful.

Susan suddenly jerked to her feet. 'The pony,' she gasped. 'Has someone remembered Falina's pony?'

She raced towards the row of stables to the right of her, fear lending wings to her feet, to a horse box that had recently been built on to the back of this wing. Would Falina's beloved pony be all right?

In the panic nobody had thought of this dumpy little animal, its screams lost in the other noises of this ghastly night, but Susan heard him as she rounded the corner, the

207

*Marquês* not far behind. Long, creeping fingers of flame were already greedily licking the roof and wooden door, billows of smoke making her cough, the heat indescribable. She was unaware of the scorching bolt as she shot it back. The pony screamed again as it leaped violently through the doorway and in its hectic rush, knocked Susan flying.

Very still she lay on the uneven paving of the small verandah, sprawled out like some rag doll abandoned by its owner.

She became half conscious and wondered where she was, her mind hazy. Oh yes, it was the awful heat of the car, that was making her feel so ill and hoped that the *quinta* wasn't much further. Peter had often spoken of the *quinta* and what was Raynor Monteiro like? Her mind wondered still further back and she was a little girl again, running in a field after her brother, shouting for him to wait for her. She never could catch Peter.

The next thing Susan knew was a tiny room with white walls, a bee buzzing at the window feeling strange and confused. Her head throbbed warningly as she moved restlessly on her pillow. Even to think was agony. Someone was sitting close beside her and as she looked at him, swift recognition came.

It was the *Marquês*, relief softening the

austere face. 'So you know me, at last,' he said grimly.

The whole event came back to her then. The fire, Nikki, Peter. 'The horses,' raising herself up.

'A little singed and very nervous, but all are safe,' he replied, easing her gently back on to her pillows.

'And the stables?' The pain in her head was torture, but she had to know.

'They were completely gutted, only the walls are left standing, but no matter, the stables can be rebuilt and this time we'll use iron roofing and not thatch. It could have been worse. And now our fair *menina* has woken up, I can go home to bed,' rubbing a hand wearily over his eyes.

'Is my hair singed?' she demanded weakly.

'A little,' a smile flitting momentarily across his mouth, 'but you are very beautiful.'

Susan didn't believe him. 'And Leonora? Is she all right? I saw her helping with the animals.'

'There is nothing wrong with her, except for a little shock. She also spent the night here and now Enrico is comforting her, getting much pleasure from it too, I'm sure.' The mocking smile was back. 'You English are far too composed,' and chuckled at her discomposure.

Susan smiled. Trust Leonora to make the most of the situation. It was the sudden realization that she would have to answer some very searching questions when she went back to the *quinta*, that made her lick her dry lips, but was grateful that the *Marquês* had not demanded an explanation now.

'You would like a drink, *senhorita*?' he asked and bent to reach for the bottle of water.

She was just about to say 'I'll manage' and stretched out for the glass, when she saw her hands for the first time. They were both bandaged.

'Your hands were slightly burned from the bolt,' he said gently, holding the glass to her lips while she drank thirstily. 'The good doctor assures me that there will be no lasting scars and now I must go.'

There was a curious tired, beaten look about him, that made Susan's heart ache, as he rose to take his leave. She was not to know that Nikki was foremost in his mind. He had spoken to that mercurial young man, but had only met with a defiant obstinacy that he had never before encountered. And why had Miss Deane gone to the stables to meet him?

The small ward was very quiet after the *Marquês* had gone, except for that wretched bee still buzzing at the window. She turned

210

restlessly. Nothing had ever happened to her in England; never been really ill, but since coming out here, she had been tipped into one silly situation after another and it had all started with that attack of heat exhaustion. If she had been a hundred percent fit, she would have seen the pitfalls and would have left the *quinta* immediately she had known that the *Marquês* knew nothing about Peter's death and on top of it all, here was the *Marquês* thinking she and Nikki were having an affair. It was all too much!

Weak tears slipped from under her eyelashes and was too tired even to wipe them away and that was how the doctor found her. He mopped her up and released that aggravating bee.

'I must be an awful nuisance,' she murmured with a half laugh.

'Not at all, *senhorita*. We all think you are a brave girl,' he replied gallantly, his fingers on her pulse. 'It is not often we have the pleasure of seeing a fair head on one of our hospital pillows.'

Two days later, Susan was allowed to go back to the *quinta* with Leonora. Enrico had come to fetch them both, but he could hardly take his eyes off his *noiva* as he tenderly helped her into the front seat of his big American car.

211

'Susan was much more hurt than I was,' she chided, but Susan could see she was extremely pleased at all the attention she was receiving.

When they arrived at the *quinta* Falina welcomed Susan as if she had been away for years. The *Marquês* was there too, smiling a greeting and then *Tia* Maria bustled out, making a great deal of fuss over Leonora, enabling Susan to slip quietly away upstairs with Falina, who said she had a dozen things to show Susan and hundreds of questions to ask. She suddenly sobered.

'Raynor told me how you saved my pony, Susan. I'm sorry you were hurt and so is Pluto.'

'Is that tubby beastie all right?'

'Only a little of his hair was crinkled and you, are you really better? It's been awful here without you.'

# 10

Some days later, Susan received a letter from Nannette in which she enthused about Alberto's wonderful house and how kind his mother was to her. Mentioned the beautiful antique furniture and his fabulous collection of old silver, the nightclubs.

Susan skimmed through the rest of the letter quickly, looking for some real news, but there was none and sighed with relief as she refolded the sheets of scented paper. Surely, if a marriage had been imminent, Nannette would have mentioned it. Her thoughts turned to Mark whom she had not seen since the party and decided to ring him to find out how he was.

When she finally managed to get through to the camp, it was to find that he was in Marula hospital with a badly crushed foot and that there was some talk that it might have to be amputated.

Anger and worry welled up inside of her as she replaced the receiver and when the *Marquês* came in for coffee a small fury confronted him.

'Why wasn't I told about Mark?' she

stormed. 'I'm not ill any more. What's a bump on the head, anyway?'

His eyes were intent as he watched her, his hands coming up in a soothing gesture.

'Miss Deane, stop it,' he ordered. 'You will make yourself ill, if you carry on like this. I'm sorry the news wasn't broken to you more gently, but the good doctor decided it was better to keep it from you for a while. You say it is nothing to have a crack on the head, but even slight concussion is something that cannot be ignored,' but Susan refused to be pacified.

'He has always been badly in need of a friend and now this — ' she gulped. 'I'm going to see him and you can't stop me,' she flung at him.

She had turned away, but the *Marquês* put a firm hand on her shoulder. '*Senhorita*, surely I'm not such an ogre. Now sit down and when you have finished your coffee, I'll take you to the hospital,' and stood over her until she had obeyed him. 'It would be the height of folly to see Fenn in your present condition. High emotion is something you do not take into a sickroom. He needs your courage, not sickly sentiment, which you would give him, in your present mood.'

'I'm sorry, *senhor*,' she apologized contritely. 'You're right. You sometimes are.'

His lips twitched. He had a shrewd suspicion as he looked down at her, that she already had the beginnings of a headache, which she would never admit to, but she had made up her mind to see Fenn immediately and he had no mind to stop her.

The drive to the hospital was covered in silence. Mark lay on his bed, his foot in an iron cage, under a white cotton bedspread.

'Oh, I am sorry,' murmured Susan as she drew up a chair, 'And I'm sorry I didn't know about your accident sooner,' and went on to tell him about the fire.

They chatted for some time, Susan appalled at the lack of response in him. It seemed as if he couldn't care less what happened to him.

'Have you heard about your foot yet?' she asked timidly. This was a Mark she had not previously known.

'Nothing as yet. If it comes off, it comes off. They make wonderful artificial limbs these days, I'm told. Every mod cons.' He smiled at her. 'You're not worried about me, are you?' he jeered.

'Mark, you're so brave!'

He turned away to gaze out of the window. 'Physically, perhaps — ' but he couldn't stand the thought of Nannette marrying someone else, thought Susan with swift understanding.

As she left the ward, she was determined to let her sister-in-law know about Mark. There would be no harm in it. Nannette would either wire back a polite message of regret or she would be up here by lunch time tomorrow, at the latest.

Susan somehow filled in the afternoon, going down to the stables with Leonora and Falina, who made a great fuss over each animal.

She shivered as she stared at the burnt out buildings, blackened and abandoned, gaping holes where the doors had been, but at the far end, workmen were busy clearing up the mess and a load of timber and roofing was stacked in one corner of the yard. This time each horse box would have a brick wall dividing it from its neighbour, Leonora informed Susan.

As they walked back to the house, Leonora asked curiously. 'I wonder how the fire started? Perhaps one of the grooms dropped a match. The hay could have smouldered for hours before bursting into flames. I suppose we'll never know.'

Nikki had been the careless one. With that thought Susan was again reminded that the *Marquês* had not yet demanded an explanation.

There was no telegram from Nannette the next morning, but about mid afternoon,

216

Susan was called to the 'phone. It was Nannette, calling from the hospital in Marula.

'Sue, I can't thank you enough for letting me know about Mark. He wouldn't have let me know.'

'Any news about his foot?' Susan was grinning widely. Her hunch had paid off.

Nannette's voice sobered. 'The doctor here, is sending him down to Beira for a second opinion tomorrow and I'll go with him. Say a prayer for him, will you? You've been such a good friend to him.'

Susan saw Mark and Nannette just before they left for Beira. It was only a hurried meeting and she did not have time to ask the hundred and one questions she was dying to know.

'Write to me, Nannette,' she urged as her sister-in-law stepped into the train, but it was a whole week before a letter arrived.

'The doctors here have decided not to amputate. Isn't that wonderful news, Sue? Mark has asked me to marry him and I've accepted and what is more, don't laugh, even said I'd be quite happy to live in a small house in Marula. We're both gloriously happy.'

Susan could hardly believe her eyes and read further. 'I'm staying at the *Grande*, as you can see from the address. I had already moved in here before I received your telegram. Suddenly I was sick to death of all the cloying adoration and felt I was just another collection piece to Alberto, to be admired and shown off to his friends.' Susan gave a deep sigh of relief and happiness as she folded up the letter.

★ ★ ★

She had just sat down to lunch the following day, when Luiz came to her and whispered that the *fidalgo* would like to see her, the *senhora* raising an eyebow, as Susan excused herself.

The *Marquês* was standing at the window, a telegram in his hand.

'It is for you, Miss Deane. Perhaps you'd like to open it here?'

Her heart jolted uncomfortably in her throat and her hands went clammy as she tore the orange envelope, then her smile radiated her relief.

'You look as if someone has left you a million *escudos*.'

'Nannette and Mark are to be married tomorrow and would like me to be there.'

'You really don't mind that Fenn is marrying *Senhora* Deane, do you,' his eyes resting on her happy mouth.

'You're invited too,' handing him the form.

'So? We'll leave here early in the morning and arrive in time for the wedding at eleven o'clock.'

How very nice it was to have all the arrangements taken out of her hands, she thought happily and then remembered.

'But I can't go alone with you, *senhor*.'

'You're learning fast, *senhorita*,' he replied smoothly, but with a distinct twinkle. 'We'll take Leonora and drop her at friends. No doubt she'll be delighted to do more shopping. Nothing seems to fascinate a woman more than buying clothes she doesn't need. Come, we'll go and tell her now.'

*Tia* Maria looked on sourly as the *Marquês* told Leonora of the proposed trip to Beira. She was delighted, even though it would mean only a few hours there, but Falina grumbled at being left behind again.

They did leave very early the following morning and by ten o'clock had arrived at the house where Leonora would be staying. *Senhora* Attunes was very happy to take Leonora shopping and Susan was able to change there too.

She had brought with her a peacock blue taffeta frock, and with her pearls and small white hat and matching accessories, hoped that she would not disgrace the bride.

Except for a swift glance of approval, the *Marquês* remained silent until they reached the city centre.

'Do you like Beira?' he asked politely and when she nodded added, 'I detect a certain reservation, Miss Deane?'

'I prefer Marula, oh, I know it's very much smaller, but it's such a friendly town.'

He shot her a quick smile. 'And your preference isn't because of a certain English gentleman either. He's getting married this morning.' He suddenly exploded. '*Meu Deus*! The man with the mark!'

Surely he wasn't taking what *ama* had said seriously? She had dismissed the old woman's words from her mind long ago. It was absurd to think, even for a minute, that Carlotta had been thinking of Nikki, even though he had that dreadful scar on the inside of his arm, when she had uttered those cryptic words? She was just about to say so when the Marques said:

'I suppose you would like to buy a gift,' and the moment was lost.

The church was dim and cool as they slipped into a pew. Susan was glad that Mark and Nannette had decided on a church wedding, even though they were the only guests, except for one other man who was standing beside Mark's wheelchair and then Nannette came down the aisle on the arm of an elderly gentleman, who, Susan found out afterwards, was the doctor.

Nannette was beautiful in a simply cut jade green suit, with a wide brimmed straw hat and a bag to match. It was the expression in those lovely eyes that put all Susan's fears to rest. They were soft and steady as they met hers. Susan was very glad that these two dear people had found each other. Tears prickled behind her closed eyelids, as the organ pealed out again, the couple coming down the aisle, Nannette walking proudly beside Mark's wheelchair, her hand curled in his.

'Hey, have a heart,' groaned Mark as Susan poured half a box of confetti over him. 'Just let the Doc here find one of the fragments in my dressing and the fat will be in the fire.'

He looked supremely happy, a different person this morning as he gazed up at his wife. Pain had etched new lines across his features, but there was also, an air of deep contentment.

'Sorry, no reception, folks. I've got to go back to the hospital. Doc here, insisted on coming with me to see that I didn't run away with my new wife,' he added with a chuckle.

'Nannette was absolutely beautiful, as a bride should be on her wedding day,' murmured Susan with satisfaction, settling herself in the comfortable seat of the car again, pulling off her hat and ruffling her hair.

'She wasn't the only beautiful one, *pequena*. Your hair reminds me of summer wheat,' the *Marquês* said with a smile.

'Please, no!' she protested, suddenly distressed and turned away from him so that only the soft curve of her cheek was visible to him.

'Why may a man not say what he pleases and I'm very much a man, Miss Deane.'

'But you're my employer.' The words came out fiercely as she turned to him again.

He was startled. 'Now here is a nicety, but one, I must admit, I failed to take into account,' he murmured and started up the car.

They covered the few kilometres back to *Senhora* Attunes' house in silence and it was only when he stopped that Susan turned to him again.

'*Senhor*, I know you must think very badly of me meeting Nikki that night, but please believe me,' her voice was shaking, 'it was not what you are thinking.' There, it was out at last.

The *Marques*' face closed and his reply was coldly polite. 'So you did go and deliberately meet him, Miss Deane?'

'Yes, he had some information I wanted.' Susan could have bitten off her tongue as soon as she had uttered the words. Fool that she was!

'About your brother?'

'Yes,' she acknowledged miserably. Now the whole story would have to come out.

'Will you tell me what it was?'

'I — I — ' she stammered, searching for the right words.

'I'm sorry, *pequena*,' he said gently, seeing her hurt, his hand coming down to cover hers. 'I found Peter always a gentleman, but — ' he shrugged, 'men do tragic things some times when they are lonely.'

Susan suddenly went limp with relief. What a reprieve! He had come to the wrong conclusion! This would keep the truth from him. The *Marquês* must never know the part Nikki had played that night in Marula. Peter was dead, there was no undoing that, but as Luiz had said and believed, the living deserved to be protected. The innocent living.

He was very gentle with her as he helped her out of the car and up the steps into the house.

★ ★ ★

They left an hour later, Leonora demanding a detailed account of the wedding and Susan was thankful for her presence on the long journey. They arrived back at the *quinta* for a late afternoon tea and after a hurried cup, the *Marquês* excused himself and disappeared into his study.

*Tia* Maria, sitting opposite Susan, watched her closely, her mouth a thin line of disapproval. To her mind Raynor was paying too much attention to this one, who, after all, was a mere companion to Leonora and Falina.

'Was the wedding a good one?' she asked after a while, curiosity getting the better of her.

'We were the only guests, apart from the bestman and the doctor, who gave Nannette away, both hospital staff, but Nannette and Mark didn't seem to mind, but it will be some time yet before he comes out of the hospital.'

The *senhora* sighed deeply, shaking her head sadly. '*Hé!* What a pity Pepita had to go back to Portugal so soon, but Raynor tells me, he has written to her father and who knows, it might not be too long before we hear of another marriage. Of course, we shall all go over to Portugal for the ceremony and perhaps there Nikki would find other distractions. *Senhorita*, you are not kind to that poor boy,' she added reprovingly. 'Why don't you go out with him to Marula or *Vila Media*? He would dearly like to show you how the fandango is danced, or perhaps you'd like to see a *torneo*?' shooting another reproachful stare at Susan, under her short fringed eyelashes. 'It has just whetted his appetite, that little rendezvous at the stables.'

Susan bit back a sharp retort. So Nikki had told her about their meeting, but what *Tia Maria* had not been told, she was sure, was the fact that it had been her beloved, spoilt Nikki who had been in that fight in Marula, over a year ago. Fond as she was of Nikki, Susan did not think she would condone such behaviour. To her the good name of the family meant as much to her as it did to the *Marquês*, the old *senhora* fiercely proud of her connection, distant though it might be. Nikki must have been mad to have done such a thing, Susan hoping the smile she pinned on her face was natural, as she changed the subject.

*Tia* Maria glared at her with dislike. There was no getting to grips with this cold girl.

' — and it's me who always has to stay at home,' finished up Falina, that night in bed. 'Why can't something nice happen to me?'

'Would you like a picnic?' suggested Susan.

'No thank you,' and then the big, dark eyes began to sparkle as her hands drew up the sheet even closer under her chin. 'Do you think I might be allowed to spend a few days with Julie?' She was Falina's very special little friend.

'I don't see why not. I think it's a lovely idea, poppet. Why don't you ask your brother tomorrow?'

The following morning both Susan and Falina were up early to pick flowers for the house. Dew still scintillated the spanned cobwebs that had to be brushed aside from the bushes.

'Why don't you go and ask your brother now if you may visit Julie?' suggested Susan, knowing the child would not settle to anything until this all important question was settled.

Falina rushed away but was soon back bubbling with excitement. She was allowed to go.

They met the *Marquês* in the hall half an hour later. 'I had a very early request this morning,' he said with a smile at the two pairs of shining eyes.

'It was kind of you, *senhor*,' murmured Susan shyly. One did not take anything for granted in this household.

He inclined his head, his glance resting on

226

Falina's basket of jumbled flowers.

'What are you going to do with those poor things?'

'Paste them in a book so that Susan will always remember our garden, but she isn't going for a long time, is she?' Her bottom lip quivered ominiously.

Her brother drew her to him. 'We'll have to see that our Miss Deane doesn't leave Mozambique,' he said gently.

Susan threw him a warning glance which he met with a bland smile. How could he deceive this little girl? Surely he must realise that she would never stay on here once he was married, or, had he manlike, not even given that a thought? If he hadn't, Pepita certainly had. How could she ever say goodbye to Falina?

The blow fell several days later. Susan was given a week's notice, the *Marquês* in his most businesslike mood, as he sat opposite her in his study. Somehow she hadn't pictured her dismissal quite like this and had just expected to fade quietly out of the picture when Miss Brack returned, but no one had even mentioned Miss Brack for some time, that almost nebulous person who had looked after Falina, before she had arrived.

Susan knew that the *Marquês* would soon be leaving for Portugal, so her dismissal now must mean that the whole family would be accompanying him, presumably to attend his wedding. Expecting all this, the shock of her notice still came as a blow. Her curious blind look made the *Marquês* throw out his hands in protest.

'Susan!' he cried, Susan!' a wealth of meaning in that single word, then hastily, as though he regretted that moment of self-revelation, rose from his chair, the old caution back.

He had never used her christian name before, she thought dully then heard him say urgently, 'This has got to be. Surely you can see that, Miss Deane?'

Yes, she realised it only too well, for Pepita certainly would not have her around when she came back here as mistress of the *quinta*. She had said so in no uncertain terms not so long ago, but a feeling of such dejection swept through her that she could only nod her head. At least the *Marquês* had sounded regretful, if that was any consolation.

'Please do not speak about this to anyone, not even to Leonora or Falina. We'll dicuss this further in a week's time.'

At the mention of Falina, a sharp anger rose in Susan. It was such a short time ago that he had said to the little girl that they would have to see that she did not leave

Mozambique. The perfidy of the man! Of course Falina must realise that she would have to leave one day, but childlike, lived only in the present, but had it been necessary to raise her hopes as he had done? Her heart ached.

Her anger gave her the courage to rise from her chair and face the *Marquês*, annoyed with herself at the tremor that had crept into her voice.

'I will do as you wish, *senhor*,' and made for the door before she made a fool of herself.

Susan quickly splashed water on her face, applied a lipstick and went down to the patio where Leonora and Falina were sitting.

'Thank you for bringing out my knitting, Falina,' she smiled and sat down on an easy chair.

'It was I who remembered,' retorted Leonora.

'Well, seeing Susan is knitting the jersey for you — ' Falina was indignant.

Here was normality again and Susan decided that it would be best to forget that she only had another week here and enjoy every minute of it, but found it wasn't that easy, for, although during the day, longings could be stifled, they surged to a new, tormenting ferocity at night and wished frantically that she could slip away and drive far into the night — to have to say goodbye to Falina and the *Marquês* —

The morning came when Falina was to spend a few days with Julie, but when the car arrived, she hung back. 'You'll still be here when I come back, won't you?' she asked anxiously.

'Of course I will, darling,' Susan replied, thankful she could, at least, be truthful. 'You're only going for a few days.'

It was during the second afternoon after Falina's departure, that Nikki drove up the long avenue of flame trees and squealed to a skidding halt in front of the house, to find Susan on her way upstairs for the *siesta*.

'Susan,' he panted, thrusting back the lock of dark hair from his high forehead. 'I've just come from *Senhora* Caldas. Falina is not at all well. Not really ill, you understand,' he added hastily, seeing her perturbed expression. 'Just a little temperature. *Senhora* Caldas suggested I take you over to see the little one for a few hours this afternoon.'

'Of course I'll go, Nikki. Thank you for coming and fetching me. Will you tell *Tia* Maria?'

At Nikki's nod she rushed upstairs and changed into a clean frock and then popped her head around Leonora's door to tell her where she was going and why.

The *senhora* was in the hall when Susan came downstairs again. 'I do not mind you going with Nikki, Susan, and I'll tell the *fidalgo* where you have gone.'

It was only when they were on the road and half way to *Vila* Media that Susan wondered why she had not been given a message from *Tia* Maria to *Senhora* Caldas, but dismissed it from her mind and began to worry about Falina. If she found her very restless, she would wrap her up warmly and take her home. Nobody could possibly take a chill in this climate, but perhaps the little girl was only homesick.

They had nearly reached *Vila* Media before Susan asked idly, 'Nikki, where does *Senhora* Caldas live?'

Instead of replying, Nikki drew up at the side of the road, with an air of a small boy caught in some naughty act.

He made a movement to take her hand which she repulsed.

'Don't be cross, Susan,' he pleaded. 'Falina isn't ill and I'm not taking you to see her.'

'Not taking me to see her?' she asked with shocked bewilderment. 'Then where are we going to?'

'I heard that Raynor had given you a week and then you would have to leave. Susan,' he was pleading now, 'do you think, once I'd felt

231

the happiness of you in my arms, I would just let you go out of my life, without doing something?'

Susan was beyond speech and he chuckled delightedly. 'Now I kidnap you. It was a bold move, was it not?' his eyes sparkling feverishly. 'It was one that does me credit to my ancestors. If they wanted a woman, they rode away with her.'

'We're not living a hundred years ago,' she reminded him tartly, fear trickling through her, but his next words reassured her somewhat.

'I only want to be allowed to spend the afternoon and evening with you, *querida*. If you had consented before, I would not have had to kidnap you. You're so cold and Nikki Monteiro is not used to that kind of treatment.'

'In your opinion, no woman should turn you down? And what do you think the *Marquês* is going to say?' she managed to ask calmly.

He drew deeply on his cigarette, flicking the ash out of the window with a quick, nervous movement, a scowl on his thin face.

'Raynor will think you have gone out with me of your own free will. After all, it is only your word against mine and I am his cousin,' he reminded her with a grin, fingering his small moustache self-consciously. 'Susan, don't let's fight,' he pleaded.

'You would never behave like this if I was a Portuguese girl. You wouldn't dare!' Anger had taken over her fear as she glared furiously at Nikki. Why hadn't she sensed that something was wrong? All her thoughts had been for Falina, away from home and ill.

'But, Susan *querida*, I was getting desperate. Soon you would be leaving and I had to see you alone, even if it was just once. I have been like a man in a fever ever since I held you in my arms at Leonora's party.'

'Don't call me *querida*,' she returned sharply. *Querida*, that lovely Portuguese word that covered so many of the tender, beautiful endearments a man kept for the one he loved. Susan had never heard it bandied about carelessly before. 'Take me home, Nikki. Immediately!'

His mouth drooped, a sure sign that he was about to sulk or be angry. 'You're one of those small girls who are delicious on the outside and hard in the centre, like some chocolates and you do not care how I feel, well, I'll show you.' His dark eyes glinted determinedly as he switched on the engine again. 'I shall take you to the bullfight whether you like it or not and then we'll dance at the Club Mozambique. You'll like that, Susan,' he said enthusiastically.

'But, Nikki — '

He held up a hand. 'You've never been to a bullfight have you, so Nikki Monteiro will take you. How can you be allowed to leave this country without having done so?'

Susan stared helplessly out of the window as they drove through the streets of *Vila Media*, thronged with people and wondered if she would be able to get a taxi back to the *quinta*.

'They're all going to the bullfight,' he said, with unusual perceptiveness, 'so don't think you'll be able to get away from me that easily.'

Susan sat and seethed, but realized that to leave him would be asking for attentions from worse than Nikki. He was harmless now, but what would he be like after a few hours at a night club? A shiver feathered her spine, but comforted herself with the thought that it would be easier to get a taxi from the club.

Nikki parked the car and they had to walk a considerable distance to the stadium, following the laughing, jostling crowd, Susan worrying uneasily about what the *Marquês'* reactions would be when *Tia* Maria told him that she had gone out with Nikki. Her eyes narrowed.

'By the way,' she demanded, 'How did you know I'd been given notice?'

He glanced possessively down at her, as she walked close by his side, pinned there by the mass of people.

'Where I get all my information from. *Tia Maria*. She tells me everything and has a fondness for me.'

She nodded, remembering seeing the *senhora* scuttling through the hall with unseemly haste that morning. She must have listened by the keyhole, for Susan felt sure the *Marquês* would not have told her of her dismissal, for he had been most emphatic that no one should know.

'You'll really enjoy the bullfight, Susan,' he assured her boyishly. 'There'll be plenty of excitement, excitement that makes the blood tingle. See if it doesn't and at the end, you'll forgive me for everything.'

Most of the seats had been taken by the time they arrived, but people good humouredly, shifted up on the long, hard benches and they were able to squeeze on to the platform.

It was not a proper bullfight. No matadors graced the scene and all and sundry could clamber into the ring to torment the bull. It was horrible and Susan cheered when the poor animal managed to retaliate with its long horns. Nikki, tense with excitement, had eyes only for the exhibition below. Suddenly he turned to her, eyes burning. 'I'm going to fight the bull, especially for you,' and before she could protest, he had pushed his way through the crowd, who cheered and clapped him as he vaulted over the railings that enclosed the arena.

Nikki must have been in the ring many times before, because he was most adept at keeping out of the animal's way. Susan sickened at the sight below. The noisy crowd, the feeling of being hemmed in, the heat and choking dust, were all becoming unbearable, making her long to get outside as soon as possible. Standing up suddenly, she determinedly pushed her way through the throng.

★ ★ ★

Back at the *quinta* the *Marquês* had returned from a visit to one of his tenants and met Leonora on the steps, looking worried.

'Oh, it's only you, Raynor,' she said with a worried frown.

'Now that wasn't very diplomatic,' he said with a smile. 'Were you expecting Enrico?'

'No, I was hoping it would be Susan. She has gone with Nikki to see Falina, who is ill, but when I asked *Tia*, all I could get out of her, was that it was nothing to worry about,' turning to go up the steps with Raynor.

'Perhaps you know what's it all about?'

Her brother shook his head and went to find the old *senhora*.

'What is this I hear about Falina?' he demanded.

'It is nothing,' but apprehension peeped out of her small, black eyes. 'It is nothing, Raynor. The little one is in good health. I have just rung *Senhora* Caldas. I will tell Timbo to bring in the tea,' but as she turned away, the *Marquês* said coldly:

'Then why has Miss Deane gone with Nikki?' he asked coldly.

'It is nothing to do with me who the *senhorita* goes out with,' bristling indignantly.

'Then why did Miss Deane tell Leonora that she was going to see Falina, who was reportedly ill?' he countered, tight lipped.

'Perhaps that one doesn't want you to know where she has gone?'

'Enough!' The *Marquês'* expression was dark with anger. 'I now want the truth.'

*Tia* Maria ducked her head deprecatingly. 'It's not so very bad, Raynor. Nikki just used a little ruse to get Susan to go out with him. I do not know where they have gone,' giving an uneasy laugh. 'The poor boy has been so downcast lately. Who does the *senhorita* think she is, to refuse a Monteiro? Miss Deane went out enough with that Englishman.'

The words had come out in an indignant gush but his angry mien made her add sullenly. 'I did not see any harm in it, Raynor. He'll look after her well.'

'And you allowed him to do such a thing?' contempt in his words.

*Tia* Maria twisted her hands in agitation. 'What does that girl mean to you, Raynor?' she demanded. 'Why do you take such an interest in her?'

His head came up arrogantly, his eyes cold and distant, but the *senhora* was too upset to see she had overstepped all bounds of what was permissable.

'Miss Deane is under my roof, therefore my responsibility and I intend to see that no harm comes to her.'

'Dear Nikki harm her? Rubbish!' It was a snort, but she seemed relieved at his words and then with a strange humbleness, asked, 'Will we hear of a betrothal soon, *Fidalgo*? This house needs a mistress and you know I only live to serve this family.'

The *Marquês* did know and because of her loyalty, he patted her shoulder and spoke more kindly. 'I certainly hope you will,' and turned to go.

She gave a satisfied smile and a nod. All would be well and that upstart of an English Miss, with her flaunting cap of gold and strange blue eyes, would soon have to go.

Luiz knocked and shuffled into the room and the *senhora* glared at him, disapprovingly, as she would have liked to have learnt more of Raynor's plans.

'Nikki has taken the *senhorita* with him,' he said without preamble. 'There is a bullfight in *Vila* Media and I'm sure he has taken her there,' ignoring *Senhora* Ferreira's gobbling noises, his anxious, sunken eyes only for the *Marquês*. 'The *senhorita* has gone thinking that Falina is ill.' It was only then that he turned to *Tia* Maria, his glance accusing.

'Is there something else I should know?' asked the *Marquês* with keen perception.

There was a pause as a frown gathered on the old man's forehead. 'I — I think you should know, *Senhor*, that *Senhor* Deane was trying to get Nikki away from a drunken brawl that night in Marula. Forgive me,' he added humbly, 'but I would not have told you, except I'm very worried about the *senhorita*. Nikki is not to be trusted.'

'What?' shrieked *Tia* Maria, her hands going up in horror. 'He would never besmirch the name of Monteiro like that! Never! Are you mad, Luiz? That girl has mesmerized you.'

He bowed. 'It is unfortunately true, *senhora*. I saw Nikki that night — '

239

The old woman's face crumpled and laboriously pressing down on her knees, with unsteady hands, lowered herself on to a chair.

The *Marquês* had remained silent, but his cold, carved expression, made Luiz want to weep.

'Did you see the whole thing?'

He shook his head. 'I only saw him running away with a horrified look about him.'

'There you are, Luiz. You saw nothing of which to accuse Nikki,' shouted the *senhora* triumphantly.

'The *senhorita* went that night of the fire to see Nikki and she told me that Nikki said the *Senhor* Deane took the blow from a heavy baton that was meant for him.'

A pregnant silence fell, broken when the *Marquês* said, 'He wouldn't harm her surely?' a curious uncertainty about him.

'Perhaps not,' conceded the old butler, 'but I would not trust that one driving a car on the return journey.'

'He won't get the opportunity. I shall find Miss Deane and bring her home myself,' and with a clasp on the old servant's trembling hands, he was gone, leaving a shattered *Tia* Maria wondering anxiously whether she would be ignominiously dismissed. As for the nonsense about Nikki, she did not believe it, shooting a vemonous scowl at Luiz. 'Why did you have to tell all those lies about *Senhor* Nikki?' she demanded angrily.

'It is the truth,' he reiterated. 'Would you like me to fetch you a glass of wine, *senhora?* You are very pale?'

*Tia* Maria ignored him, rising majestically to her feet and after an angry stare, swept past the butler, her head held high.

$$\star \quad \star \quad \star$$

Susan, pushing her way through the mass of humanity, found herself going in the opposite direction to that of the crowds still pouring into the stadium. One huge man, with glittering coal-black eyes and massive shoulders, took her lightly by the waist and swinging her up against him, gave her a resounding kiss. The people around them laughed delightedly and clapped their approval. It was all part of the day's entertainment, although some of them did wonder curiously at the presence of this fair haired girl in their midst.

Susan was furious and the crowd was stilled as she gave the man a stinging slap on the cheek before he put her down. There was another outburst of laughter at the big man's obvious discomfort.

She realised, too late, that it would have

been far better if she had also laughed off the whole incident and let it pass, for suddenly the mood of the people changed and she found herself surrounded by an angry, gesticulating mob, shouting insults.

Panic rose chokingly within Susan and wished fervently, that she hadn't been so hasty in trying to get away from Nikki. She gazed around her like a trapped animal for a way through, but only encountered a wall of hostile faces and then, miraculously, there stood the *Marquês*.

His hand came up hard under her arm and the crowd turned away. They all knew Raynor Monteiro.

Susan's eyes widened in disbelief and then collapsed against him. Never, never had she been so thankful to see anyone!

'So, you cannot always look after yourself, my independent Miss Deane,' he observed blandly as he marched her out of the stadium and she noticed, with a strange breathlessness, that he had been really worried about her. It had roughed his voice and she wondered if he would give rein to his temper and shake her, but they walked in silence to where his car was parked in a nearby garage.

Thankfully, Susan dropped on to the front seat and when the *Marquês* slipped behind the wheel she asked a little hesitantly, 'Don't you think you should tell Nikki that you are taking me home?'

'No, I don't, even if I could find him in all that crowd. I hope he spends a couple of hellish hours looking for you. It will teach him a lesson.'

She glanced uneasily at him and saw his hands tightly clenched on the wheel, the knuckles gleaming white and could feel the mounting tension fill the car. Her voice faltered as she said, 'You don't know how thankful I was to see you, *senhor*! It was horrible! That poor bull and the crowds!' and felt her hands trembling. Reaction had set in, leaving her fighting tears that threatened to overflow.

The *Marquês* drew up at the side of the road and stopped, opened the glove compartment and drew out a small flask, pouring the amber liquid into the top. She knew it was useless to protest for he was quite capable of forcing it down her throat, but the cup clattered so much against her teeth that she couldn't swallow. Taking the cup from her, he held it, putting his arm close around her.

'Now let's have none of this stiff upper lip nonsense,' he said briskly. 'Let the tears come, little one. All this tension is bad.'

Strangely enough, no tears came and her shuddering ceased. 'Thank you, *senhor*,' in the most dignified way she could muster and drew away from him.

He started up the car and they drove swiftly out of town, leaving the crowds and the noise behind, the road stretching before them, silent and empty, the only living creature, an old man leading a well laden donkey along a distant footpath.

Susan was still expecting his anger to break as he led her up the steps into the *sala*, but none came and was surprised when he said sombrely:

'*Senhorita*, you have much to forgive this family,' seating himself in a chair next to her. 'I shall send Nikki back to Portugal,' and she knew that Leonora must have told him what had occurred.

'Don't blame Nikki too much, *senhor*. He is very young,' thinking of what this would mean to *Tia* Maria. It would break her heart.

'You would plead for him after what he did this afternoon?'

A brooding silence fell and she glanced anxiously at the man beside her. Surely Nikki taking her to the bullfight shouldn't affect him like this? He seemed crushed under a weight too heavy to bear, but that was nonsense. This was the *Marquês* and had never seen him like this before. Had he received bad news?

'It was an unpardonable thing to have done, to leave you in that crowd by yourself,' a flash of anger lighting his dark eyes for a moment. 'Not to mention forcing you to go with him by despicable means. No, I've had enough of that young man.' His expression was cold, implacable, his words final. 'Nikki will go back and work for his father in their wine exporting business. I'd also like to thank you for keeping from me Nikki's part in the affair in which Peter was killed. You went to meet him, that night of the fire, to find out this, not so?'

Susan nodded. Who had told him, she wondered. It could only have been Luiz.

He was leaning forward, his hands hanging between his knees, his eyes intent on the match between his fingers. 'What must you think of my family,' he muttered. 'If it had not been for Nikki, your brother would have been alive today, Luiz told me.'

She longed to be able to comfort him, longed to hold him close to soothe away this deep hurt she knew was tormenting him, but it was not her place.

'*Senhor*, no one will ever hear about it from me,' she said gently.

'I'll live with it all my days,' he replied slowly. 'That one of my family should have sunk so low as to be mixed up in a common fight and now this.'

'But I am here, safe and none the worse for my experience. Thank you, *senhor*, for coming to my rescue and I can always boast, when I get back to England, that I did witness a bullfight,' she ended up lightly.

'And if I hadn't come?' He glanced up swiftly.

She was silenced as that awful scene rushed back. Her terrified reaction to that hostile crowd, those poor bulls!

After dinner that night, a subdued and unusually incoherent *senhora* came to see Susan.

'*Senhorita*, please forgive me,' tears running down her cheeks. 'I didn't know, I — I didn't guess. Nikki — ' She stopped as if afraid she had already said too much, adding humbly, 'I'm to stay.'

'I'm glad, *Tia* Maria,' Susan replied gently, 'for what would this family do without you? Who would run this huge house as ably as you do?'

Gratified, a little of the tension disappeared as the *senhora* nodded and a trembling smile broke her woeful expression. 'The *senhorita* is most kind. Thank you, Susan, but Nikki?'

'Perhaps the *Marquês* will relent?'

'No!' The word came out explosively and Susan stared at her in surprise. 'Nikki, even before you came here, was worrying Raynor. It is better that he go home. He is very headstrong, you understand,' excusing him.

★   ★   ★

Two mornings later, Susan awoke with a dull ache around her heart. This could well be her last day here, for her week's notice had terminated the night before. She hadn't slept well, either, thinking of her parting with Falina. The little girl had arrived back at the *quinta* after lunch yesterday and had been glad to be home again.

She went slowly downstairs, imprinting every little detail on her mind. The wide, curved staircase, the huge hall below where she had arranged so many flowers, the inner garden, where she and Falina had spent many happy hours. The vast ballroom which, no doubt, would be opened again for Leonora's wedding.

Susan murmured a hasty apology for being late as she slipped into the chair the *Marquês* held out for her.

He was at his most affable, she noticed and there was even a humanizing twinkle in his eyes. The *senhora*, on the other hand, appeared ill.

'*Senhora*,' Susan said impulsively, 'wouldn't

247

you like to visit your sister this morning? I'll see to things here.'

'It is very kind of you, *senhorita*,' turning to the *Marquês* for approval.

'You may go tomorrow, *Tia* Maria. I promised *Ama* Carlotta I would take Miss Deane to see her this morning.'

'May we go too?' chimed in Falina eagerly, but her brother shook his head.

The two girls excused themselves and left the room, but as they went along the passage, Falina's shrill, indignant treble could easily be heard as she demanded to know why Leonora had kicked her.

The *Marquês'* mouth twitched, but all he said was, 'Can you be ready in half an hour's time, Miss Deane?'

The *senhora* was shocked, her mouth open to expostulate, but a quick stare from the *fidalgo* made her subside with a few dark mutterings.

Susan was waiting on the patio as the *Marquês* arrived with the car, cool in a candy-striped frock in blue, pink and white accompanied by a wide brimmed straw hat and ran down the shallow steps to meet him just as he was opening the door for her.

They drove down the avenue in silence, the sun etching patches of light and dark on the sandy surface as the trees flashed by. Susan in her corner, felt lost this morning, but was glad of this opportunity to say goodbye to Carlotta and wondered if the old woman had been told that she would be going back to England.

The car slowed down and she glanced across enquiringly at the *Marquês* as he drew to the side of the road and stopped under a big leafy tree and was even more puzzled when he said with a quiet purpose:

'Now, perhaps, we can talk without any interruptions,' and switched off the engine. 'I promised *Ama* I would, before I took you to see her this morning.' He had half turned to her, drinking in every line of her face, or so it seemed to Susan, who felt the treacherous colour sweep under her fair skin.

He had never looked at her like that before, as if he were offering her his heart. A heart that belonged to Pepita, she reminded herself, drawing further into her corner, with a small protesting movement.

'My week is up, *senhor*. I suppose that is what you wish to talk about? I'd like to leave early this afternoon,' she said with quiet dignity, turning to stare out of the window, her cheeks still hot, her hands clenched in her lap.

'But that is not the reason why I gave you notice, Miss Deane,' he said quietly. 'You reminded me, that employers should not, in your estimation, pay compliments to their staff, but I'm no longer your employer, so now I may talk to you as a man to a woman, a man who loves you very much. Will you marry me, little one?' he asked simply, the words low, without a trace of his old arrogant smoothness, which he had so often adopted.

Susan sat very still, blinking in surprise. He had taken her hands in his and was gently opening her fingers. Soon, she thought, this delicate, fragile ball of happiness within her would smash into a thousand fragments and she would awake to find this to be just another of her foolish dreams. The *Marquês* asking her to marry him? The Marquês?

Jolted back to her senses, she sat up with a gasp. 'Oh, no, *senhor!*' shocked reproof in the words.

'And why not?' he enquired patiently, his eyes darkly tender as they lingered on her perturbed face. 'I give you all my heart, *querida*. I have loved you from that moment I carried you upstairs and knew then I had to try and keep you at the *quinta* at all costs, or else I would lose you. You will never know what a cruel stab it was when I thought you were Peter's wife. I had thought for some time that she should have been out here with him.'

'But, I'm just an ordinary person,' she said in bewilderment. 'You can't marry me and what about Pepita? I thought your marriage had all been arranged.'

Instead of the expressive eyebrows coming down in two haughty lines, his mouth quirked a little wryly at the corners.

'I won't hide anything from you, *pequena*. I was contemplating marriage. Yes, an arranged marriage, you so much despise, but you must understand,' he added quickly, 'Pepita only came out here for a visit. She did not know that any arrangements had been made and very airy fairy arrangements they were too.' He shrugged. 'She might have wondered, but *Senhora* Oliveira and Pepita had visited many families here in Mozambique, all with marriageable sons, but then I rescued a fair English *menina* on the road to the *quinta* and I knew, with a thousand certainties, that I could never contemplate such an alliance. Well, Susan, you're not usually so silent. Will you be my wife?' with a tender smile, drawing her gently towards him.

'You, the *fidalgo*, should know the answer,' she murmured, suddenly shy, her face hidden in his cream shirt, feeling the warmth and the beat of his heart against her cheek.

'One does not always have the courage to presume, *querida*. My love, tell me,' he

251

insisted, somewhat in his old manner.

'But, senhor — ' No, she would not tell him yet. There were other matters to be straightened out first.

'It will never be 'senhor' again, but what is worrying you, querida?' in a tone that melded patience with a gentle mockery. 'Surely, it is not an unsurmountable barrier?'

'I think you feel in some way responsible for me, because of Peter, but you needn't,' her eyes though, were troubled, as she tried, unsuccessfully, to pull herself out of his arms.

'How can I make one small, determined girl believe that I love her for herself alone, hé?' he muttered into the soft silkiness of her hair. There was nothing impersonal this time, about his fingers caressing the back of her neck.

Susan could feel his hands, so tender and yet with an urgency that thrilled her and she knew she could no longer deny him her answer.

'Yes, oh, yes, I do love you,' she whispered softly.

For a moment she was conscious of the relaxing of his body against her as if he hadn't been at all sure of her answer. He drew her even closer into his arms and as he tilted her face, found her lips, cool and sweet and sealed them with his own.

Some time later, her head on his shoulder, Susan said tremulously. 'I still think I shouldn't marry you. What will your family say?'

'Enough! Please do not put any more obstacles in my path, Susan.'

'You've only twice before used my christian name. I've wondered about that.'

'I felt I hadn't the right,' he replied simply, raising her hand to his lips. 'There seemed no hope for me with Fenn on the scene and even, as I thought Nikki.' He chuckled. 'I eventually took my troubles to my old *Ama*.'

'But she was wrong, saying I would marry the man with the mark. Not that I believed her,' she added hastily.

'But she was right, *querida*. Look.' He opened his shirt and there, high on his shoulder was a birthmark.

Susan's eyes widened. 'Carlotta didn't mean a scar. It was you she was referring to,' she breathed. '*Ama* knew, even then?'

'Yes, I also thought she meant that and it made me furiously angry, and jealous. Nikki would never have made you content. He has the roving eye, that one,' chuckling added. 'And it has been very difficult to try and court you under the suspicious, twitching nose of *Tia Maria*.'

'And don't forget the conventions,' she added happily.

'And the conventions,' he agreed. 'We'll spend our honeymoon on a desert island where there will be no relations or even well-meaning friends,' but Susan knew that he would take her to Portugal.

'I have spent the most unhappy week of my life, thinking I'd have to leave the *quinta*,' she couldn't help scolding, but with a warm tenderness in her eyes.

'I'm sorry, little one.' He was so contrite that it was hard to realize this really was the *Marquês*. 'I'll make it up to you for the rest of my life, *querida*. Do you know I very nearly told you of my love then? I've never seen such hurt on anyone's face, but it led me to hope that perhaps you did love me, but come, I've promised *Ama* I would bring you to her as soon as I'd asked you to marry me.'

He bent and kissed her again and this time she returned his caresses with a fervour that matched his own. 'You learn fast, sweetheart,' he murmured tenderly. 'It would have broken my heart if I'd found you cold,' and kissed her again.

She dropped her gaze from his ardent one, a slow tide of colour suffusing her cheeks as he straightened her fringe with his fingers. 'I have often longed to smooth down these ruffled feathers of yours, *querida*.'

The *Marquês* started up the car and as they turned a corner, the small cottage came into view and there on the verandah sat Carlotta, patiently waiting for them, dressed in her *fiesta* finery.

Susan chuckled. How very sure the old *ama* had been that her beloved *fidalgo* would find the happiness he had been seeking for so long.

## THE END

We do hope that you have enjoyed reading this large print book.

Did you know that all of our titles are available for purchase?

We publish a wide range of high quality large print books including:
**Romances, Mysteries, Classics, General Fiction, Non Fiction and Westerns.**

Special interest titles available in large print are:
**The Little Oxford Dictionary**
**Music Book**
**Song Book**
**Hymn Book**
**Service Book**

Also available from us courtesy of Oxford University Press:
**Young Readers' Dictionary**
**(large print edition)**
**Young Readers' Thesaurus**
**(large print edition)**

For further information or a free brochure, please contact us at:
**Ulverscroft Large Print Books Ltd., The Green, Bradgate Road, Anstey, Leicester, LE7 7FU, England.**
**Tel:** (00 44) 0116 236 4325
**Fax:** (00 44) 0116 234 0205

# STANDING IN THE SHADOWS

## Michelle Spring

Laura Principal is repelled but fascinated as she investigates the case of an eleven-year-old boy who has murdered his foster mother. It is not the sort of crime one would expect in Cambridge. The child, Daryll, has confessed to the brutal killing; now his elder brother wants to find out what has turned him into a ruthless killer. Laura confronts an investigation which is increasingly tainted with violence. And that's not all. Someone with an interest in the foster mother's murder is standing in the shadows, watching her every move . . .